In Praise of The RISC-V Reader

This timely book concisely describes the simple, free and open RISC-V ISA that is experiencing rapid uptake in many different computing sectors. The book also contains many insights about computer architecture in general, as well as explaining the particular design choices we made in creating RISC-V. I can imagine this book becoming a well-worn reference guide for many RISC-V practitioners.

—Professor Krste Asanović, University of California, Berkeley, one of the four architects of RISC-V

I like RISC-V and this book as they are elegant—brief, to the point, and complete. The book's commentaries provide a gratuitous history, motivation, and architecture critique.

—C. Gordon Bell, Microsoft and designer of the Digital PDP-11 and VAX-11 instruction set architectures

This handy little book effortlessly summarizes all the essential elements of the RISC-V Instruction Set Architecture, a perfect reference guide for students and practitioners alike.

—Professor Randy Katz, University of California, Berkeley, one of the inventors of RAID storage systems

RISC-V is a fine choice for students to learn about instruction set architecture and assembly-level programming, the basic underpinnings for later work in higher-level languages. This clearly-written book offers a good introduction to RISC-V, augmented with insightful comments on its evolutionary history and comparisons with other familiar architectures. Drawing on past experience with other architectures, RISC-V designers were able avoid unnecessary, often irregular features, yielding easy pedagogy. Although simple, it is still powerful enough for widespread use in real applications. Long ago, I used to teach a first course in assembly programming and if I were doing that now, I'd happily use this book.

—John Mashey, one of the designers of the MIPS instruction set architecture

This book tells what RISC-V can do and why its designers chose to endow it with those abilities. Even more interesting, the authors tell why RISC-V omits things found in earlier machines. The reasons are at least as interesting as RISC-V's endowments and omissions.

—Ivan Sutherland, Turing Award laureate called the father of computer graphics

RISC-V will change the world, and this book will help you become part of that change.

—Professor Michael B. Taylor, University of Washington

This book will be an invaluable reference for anyone working with the RISC-V ISA. The opcodes are presented in several useful formats for quick reference, making assembly coding and interpretation easy. In addition, the explanations and examples of how to use the ISA make the programmer's job even simpler. The comparisons with other ISAs are interesting and demonstrate why the RISC-V creators made the design decisions they did.

—Megan Wachs, PhD, SiFive Engineer

Open ◢◤RISC-V Reference Card ①

Base Integer Instructions: RV32I and RV64I

Category	Name	Fmt	RV32I Base	+RV64I
Shifts Shift Left Logical		R	SLL rd,rs1,rs2	SLLW rd,rs1,rs2
Shift Left Log. Imm.		I	SLLI rd,rs1,shamt	SLLIW rd,rs1,shamt
Shift Right Logical		R	SRL rd,rs1,rs2	SRLW rd,rs1,rs2
Shift Right Log. Imm.		I	SRLI rd,rs1,shamt	SRLIW rd,rs1,shamt
Shift Right Arithmetic		R	SRA rd,rs1,rs2	SRAW rd,rs1,rs2
Shift Right Arith. Imm.		I	SRAI rd,rs1,shamt	SRAIW rd,rs1,shamt
Arithmetic ADD		R	ADD rd,rs1,rs2	ADDW rd,rs1,rs2
ADD Immediate		I	ADDI rd,rs1,imm	ADDIW rd,rs1,imm
SUBtract		R	SUB rd,rs1,rs2	SUBW rd,rs1,rs2
Load Upper Imm		U	LUI rd,imm	
Add Upper Imm to PC		U	AUIPC rd,imm	
Logical XOR		R	XOR rd,rs1,rs2	
XOR Immediate		I	XORI rd,rs1,imm	
OR		R	OR rd,rs1,rs2	
OR Immediate		I	ORI rd,rs1,imm	
AND		R	AND rd,rs1,rs2	
AND Immediate		I	ANDI rd,rs1,imm	
Compare Set <		R	SLT rd,rs1,rs2	
Set < Immediate		I	SLTI rd,rs1,imm	
Set < Unsigned		R	SLTU rd,rs1,rs2	
Set < Imm Unsigned		I	SLTIU rd,rs1,imm	
Branches Branch =		B	BEQ rs1,rs2,imm	
Branch ≠		B	BNE rs1,rs2,imm	
Branch <		B	BLT rs1,rs2,imm	
Branch ≥		B	BGE rs1,rs2,imm	
Branch < Unsigned		B	BLTU rs1,rs2,imm	
Branch ≥ Unsigned		B	BGEU rs1,rs2,imm	
Jump & Link J&L		J	JAL rd,imm	
Jump & Link Register		I	JALR rd,rs1,imm	
Synch Synch thread		I	FENCE	
Synch Instr & Data		I	FENCE.I	
Environment CALL		I	ECALL	
BREAK		I	EBREAK	

Control Status Register (CSR)

Category	Name	Fmt	RV32I Base
Read/Write		I	CSRRW rd,csr,rs1
Read & Set Bit		I	CSRRS rd,csr,rs1
Read & Clear Bit		I	CSRRC rd,csr,rs1
Read/Write Imm		I	CSRRWI rd,csr,imm
Read & Set Bit Imm		I	CSRRSI rd,csr,imm
Read & Clear Bit Imm		I	CSRRCI rd,csr,imm

Category	Name	Fmt	RV32I Base
Loads Load Byte		I	LB rd,rs1,imm
Load Halfword		I	LH rd,rs1,imm
Load Byte Unsigned		I	LBU rd,rs1,imm
Load Half Unsigned		I	LHU rd,rs1,imm
Load Word		I	LW rd,rs1,imm
Stores Store Byte		S	SB rs1,rs2,imm
Store Halfword		S	SH rs1,rs2,imm
Store Word		S	SW rs1,rs2,imm

+RV64I	
LWU rd,rs1,imm	
LD rd,rs1,imm	
SD rs1,rs2,imm	

RV Privileged Instructions

Category	Name	Fmt	RV mnemonic
Trap Mach-mode trap return		R	MRET
Supervisor-mode trap return		R	SRET
Interrupt Wait for Interrupt		R	WFI
MMU Virtual Memory FENCE		R	SFENCE.VMA rs1,rs2

Examples of the 60 RV Pseudoinstructions

		Fmt	
Branch = 0 (BEQ rs,x0,imm)		J	BEQZ rs,imm
Jump (uses JAL x0,imm)		J	J imm
MoVe (uses ADDI rd,rs,0)		R	MV rd,rs
RETurn (uses JALR x0,0,ra)		I	RET

Optional Compressed (16-bit) Instruction Extension: RV32C

Category	Name	Fmt	RVC	RISC-V equivalent
Loads Load Word		CL	C.LW rd',rs1',imm	LW rd',rs1',imm*4
Load Word SP		CI	C.LWSP rd,imm	LW rd,sp,imm*4
Float Load Word SP		CL	C.FLW rd',rs1',imm	FLW rd',rs1',imm*8
Float Load Word		CI	C.FLWSP rd,imm	FLW rd,sp,imm*8
Float Load Double		CL	C.FLD rd',rs1',imm	FLD rd',rs1',imm*16
Float Load Double SP		CI	C.FLDSP rd,imm	FLD rd,sp,imm*16
Stores Store Word		CS	C.SW rs1',rs2',imm	SW rs1',rs2',imm*4
Store Word SP		CSS	C.SWSP rs2,imm	SW rs2,sp,imm*4
Float Store Word		CS	C.FSW rs1',rs2',imm	FSW rs1',rs2',imm*8
Float Store Word SP		CSS	C.FSWSP rs2,imm	FSW rs2,sp,imm*8
Float Store Double		CS	C.FSD rs1',rs2',imm	FSD rs1',rs2',imm*16
Float Store Double SP		CSS	C.FSDSP rs2,imm	FSD rs2,sp,imm*16
Arithmetic ADD		CR	C.ADD rd,rs1	ADD rd,rd,rs1
ADD Immediate		CI	C.ADDI rd,imm	ADDI rd,rd,imm
ADD SP Imm * 16		CI	C.ADDI16SP x0,imm	ADDI sp,sp,imm*16
ADD SP Imm * 4		CIW	C.ADDI4SPN rd',imm	ADDI rd',sp,imm*4
SUB		CR	C.SUB rd,rs1	SUB rd,rd,rs1
AND		CR	C.AND rd,rs1	AND rd,rd,rs1
AND Immediate		CI	C.ANDI rd,imm	ANDI rd,rd,imm
OR		CR	C.OR rd,rs1	OR rd,rd,rs1
eXclusive OR		CR	C.XOR rd,rs1	AND rd,rd,rs1
MoVe		CR	C.MV rd,rs1	ADD rd,rs1,x0
Load Immediate		CI	C.LI rd,imm	ADDI rd,x0,imm
Load Upper Imm		CI	C.LUI rd,imm	LUI rd,imm
Shifts Shift Left Imm		CI	C.SLLI rd,imm	SLLI rd,rd,imm
Shift Right Ari. Imm.		CI	C.SRAI rd,imm	SRAI rd,rd,imm
Shift Right Log. Imm.		CI	C.SRLI rd,imm	SRLI rd,rd,imm
Branches Branch=0		CB	C.BEQZ rs1',imm	BEQ rs1',x0,imm
Branch≠0		CB	C.BNEZ rs1',imm	BNE rs1',x0,imm
Jump Jump		CJ	C.J imm	JAL x0,imm
Jump Register		CR	C.JR rd,rs1	JALR x0,rs1,0
Jump & Link J&L		CJ	C.JAL imm	JAL ra,imm
Jump & Link Register		CR	C.JALR rs1	JALR ra,rs1,0
System Env. BREAK		CI	C.EBREAK	EBREAK

Optional Compressed Extention: RV64C

All RV32C (except C.JAL, 4 word loads, 4 word strores) plus:

ADD Word (C.ADDW)			Load Doubleword (C.LD)
ADD Imm. Word (C.ADDIW)			Load Doubleword SP (C.LDSP)
SUBtract Word (C.SUBW)			Store Doubleword (C.SD)
			Store Doubleword SP (C.SDSP)

32-bit Instruction Formats

	31 27	26 25	24 20	19 15	14 12	11 7	6 0
R	funct7		rs2	rs1	funct3	rd	opcode
I	imm[11:0]			rs1	funct3	rd	opcode
S	imm[11:5]		rs2	rs1	funct3	imm[4:0]	opcode
B	imm[12|10:5]		rs2	rs1	funct3	imm[4:1|11]	opcode
U	imm[31:12]					rd	opcode
J	imm[20|10:1|11|19:12]					rd	opcode

16-bit (RVC) Instruction Formats

	15 14 13	12	11 10 9	8 7	6 5	4 3 2	1 0
CR	funct4		rd/rs1		rs2		op
CI	funct3	imm	rd/rs1		imm		op
CSS	funct3		imm			rs2	op
CIW	funct3		imm			rd'	op
CL	funct3	imm	rs1'		imm	rd'	op
CS	funct3	offset	rs1'		offset	rs2'	op
CB	funct3	offset	rs1'		offset		op
CJ	funct3		jump target				op

RISC-V Integer Base (RV32I/64I), privileged, and optional RV32/64C. Registers x1–x31 and the PC are 32 bits wide in RV32I and 64 in RV64I (x0=0). RV64I adds 12 instructions for the wider data. Every 16-bit RVC instruction maps to an existing 32-bit RISC-V instruction.

Optional Multiply-Divide Instruction Extension: RVM

Category	Name	Fmt	RV32M (Multiply-Divide)		+RV64M	
Multiply	MULtiply	R	MUL	rd,rs1,rs2	MULW	rd,rs1,rs2
	MULtiply High	R	MULH	rd,rs1,rs2		
	MULtiply High Sign/Uns	R	MULHSU	rd,rs1,rs2		
	MULtiply High Uns	R	MULHU	rd,rs1,rs2		
Divide	DIVide	R	DIV	rd,rs1,rs2	DIVW	rd,rs1,rs2
	DIVide Unsigned	R	DIVU	rd,rs1,rs2		
Remainder	REMainder	R	REM	rd,rs1,rs2	REMW	rd,rs1,rs2
	REMainder Unsigned	R	REMU	rd,rs1,rs2	REMUW	rd,rs1,rs2

Optional Atomic Instruction Extension: RVA

Category	Name	Fmt	RV32A (Atomic)		+RV64A	
Load	Load Reserved	R	LR.W	rd,rs1	LR.D	rd,rs1
Store	Store Conditional	R	SC.W	rd,rs1,rs2	SC.D	rd,rs1,rs2
Swap	SWAP	R	AMOSWAP.W	rd,rs1,rs2	AMOSWAP.D	rd,rs1,rs2
Add	ADD	R	AMOADD.W	rd,rs1,rs2	AMOADD.D	rd,rs1,rs2
Logical	XOR	R	AMOXOR.W	rd,rs1,rs2	AMOXOR.D	rd,rs1,rs2
	AND	R	AMOAND.W	rd,rs1,rs2	AMOAND.D	rd,rs1,rs2
	OR	R	AMOOR.W	rd,rs1,rs2	AMOOR.D	rd,rs1,rs2
Min/Max	MINimum	R	AMOMIN.W	rd,rs1,rs2	AMOMIN.D	rd,rs1,rs2
	MAXimum	R	AMOMAX.W	rd,rs1,rs2	AMOMAX.D	rd,rs1,rs2
	MINimum Unsigned	R	AMOMINU.W	rd,rs1,rs2	AMOMINU.D	rd,rs1,rs2
	MAXimum Unsigned	R	AMOMAXU.W	rd,rs1,rs2	AMOMAXU.D	rd,rs1,rs2

Two Optional Floating-Point Instruction Extensions: RVF & RVD

Category	Name	Fmt	RV32{F|D} (SP,DP Fl. Pt.)		+RV64{F|D}	
Move	Move from Integer	R	FMV.W.X	rd,rs1	FMV.D.X	rd,rs1
	Move to Integer	R	FMV.X.W	rd,rs1	FMV.X.D	rd,rs1
Convert	ConVerT from Int	R	FCVT.{S|D}.W	rd,rs1	FCVT.{S|D}.L	rd,rs1
	ConVerT from Int Unsigned	R	FCVT.{S|D}.WU	rd,rs1	FCVT.{S|D}.LU	rd,rs1
	ConVerT to Int	R	FCVT.W.{S|D}	rd,rs1	FCVT.L.{S|D}	rd,rs1
	ConVerT to Int Unsigned	R	FCVT.WU.{S|D}	rd,rs1	FCVT.LU.{S|D}	rd,rs1
Load	Load	I	FL{W,D}	rd,rs1,imm		
Store	Store	S	FS{W,D}	rs1,rs2,imm		
Arithmetic	ADD	R	FADD.{S|D}	rd,rs1,rs2		
	SUBtract	R	FSUB.{S|D}	rd,rs1,rs2		
	MULtiply	R	FMUL.{S|D}	rd,rs1,rs2		
	DIVide	R	FDIV.{S|D}	rd,rs1,rs2		
	SQuare RooT	R	FSQRT.{S|D}	rd,rs1		
Mul-Add	Multiply-ADD	R	FMADD.{S|D}	rd,rs1,rs2,rs3		
	Multiply-SUBtract	R	FMSUB.{S|D}	rd,rs1,rs2,rs3		
	Negative Multiply-SUBtract	R	FNMSUB.{S|D}	rd,rs1,rs2,rs3		
	Negative Multiply-ADD	R	FNMADD.{S|D}	rd,rs1,rs2,rs3		
Sign Inject	SiGN source	R	FSGNJ.{S|D}	rd,rs1,rs2		
	Negative SiGN source	R	FSGNJN.{S|D}	rd,rs1,rs2		
	Xor SiGN source	R	FSGNJX.{S|D}	rd,rs1,rs2		
Min/Max	MINimum	R	FMIN.{S|D}	rd,rs1,rs2		
	MAXimum	R	FMAX.{S|D}	rd,rs1,rs2		
Compare	compare Float =	R	FEQ.{S|D}	rd,rs1,rs2		
	compare Float <	R	FLT.{S|D}	rd,rs1,rs2		
	compare Float ≤	R	FLE.{S|D}	rd,rs1,rs2		
Categorize	CLASSify type	R	FCLASS.{S|D}	rd,rs1		
Configure	Read Status	R	FRCSR	rd		
	Read Rounding Mode	R	FRRM	rd		
	Read Flags	R	FRFLAGS	rd		
	Swap Status Reg	R	FSCSR	rd,rs1		
	Swap Rounding Mode	R	FSRM	rd,rs1		
	Swap Flags	R	FSFLAGS	rd,rs1		
	Swap Rounding Mode Imm	I	FSRMI	rd,imm		
	Swap Flags Imm	I	FSFLAGSI	rd,imm		

Optional Vector Extension: RVV

Name	Fmt	RV32V/R64V	
SET Vector Len.	R	SETVL	rd,rs1
MULtiply High	R	VMULH	rd,rs1,rs2
REMainder	R	VREM	rd,rs1,rs2
Shift Left Log.	R	VSLL	rd,rs1,rs2
Shift Right Log.	R	VSRL	rd,rs1,rs2
Shift R. Arith.	R	VSRA	rd,rs1,rs2
LoaD	I	VLD	rd,rs1,imm
LoaD Strided	R	VLDS	rd,rs1,rs2
LoaD indeXed	R	VLDX	rd,rs1,rs2
STore	S	VST	rd,rs1,imm
STore Strided	R	VSTS	rd,rs1,rs2
STore indeXed	R	VSTX	rd,rs1,rs2
AMO SWAP	R	AMOSWAP	rd,rs1,rs2
AMO ADD	R	AMOADD	rd,rs1,rs2
AMO XOR	R	AMOXOR	rd,rs1,rs2
AMO AND	R	AMOAND	rd,rs1,rs2
AMO OR	R	AMOOR	rd,rs1,rs2
AMO MINimum	R	AMOMIN	rd,rs1,rs2
AMO MAXimum	R	AMOMAX	rd,rs1,rs2
Predicate =	R	VPEQ	rd,rs1,rs2
Predicate ≠	R	VPNE	rd,rs1,rs2
Predicate <	R	VPLT	rd,rs1,rs2
Predicate ≥	R	VPGE	rd,rs1,rs2
Predicate AND	R	VPAND	rd,rs1,rs2
Pred. AND NOT	R	VPANDN	rd,rs1,rs2
Predicate OR	R	VPOR	rd,rs1,rs2
Predicate XOR	R	VPXOR	rd,rs1,rs2
Predicate NOT	R	VPNOT	rd,rs1
Pred. SWAP	R	VPSWAP	rd,rs1
MOVe	R	VMOV	rd,rs1
ConVerT	R	VCVT	rd,rs1
ADD	R	VADD	rd,rs1,rs2
SUBtract	R	VSUB	rd,rs1,rs2
MULtiply	R	VMUL	rd,rs1,rs2
DIVide	R	VDIV	rd,rs1,rs2
SQuare RooT	R	VSQRT	rd,rs1,rs2
Multiply-ADD	R	VFMADD	rd,rs1,rs2,rs3
Multiply-SUB	R	VFMSUB	rd,rs1,rs2,rs3
Neg. Mul.-SUB	R	VFNMSUB	rd,rs1,rs2,rs3
Neg. Mul.-ADD	R	VFNMADD	rd,rs1,rs2,rs3
SiGN inJect	R	VSGNJ	rd,rs1,rs2
Neg SiGN inJect	R	VSGNJN	rd,rs1,rs2
Xor SiGN inJect	R	VSGNJX	rd,rs1,rs2
MINimum	R	VMIN	rd,rs1,rs2
MAXimum	R	VMAX	rd,rs1,rs2
XOR	R	VXOR	rd,rs1,rs2
OR	R	VOR	rd,rs1,rs2
AND	R	VAND	rd,rs1,rs2
CLASS	R	VCLASS	rd,rs1
SET Data Conf.	R	VSETDCFG	rd,rs1
EXTRACT	R	VEXTRACT	rd,rs1,rs2
MERGE	R	VMERGE	rd,rs1,rs2
SELECT	R	VSELECT	rd,rs1,rs2

Calling Convention

Register	ABI Name	Saver	Description
x0	zero	---	Hardwired zero
x1	ra	Caller	Return address
x2	sp	Callee	Stack pointer
x3	gp	---	Global pointer
x4	tp	---	Thread pointer
x5-7	t0-2	Caller	Temporaries
x8	s0/fp	Callee	Saved registers
x9	s1	Callee	Function args
x10-11	a0-1	Caller	
x12-17	a2-7	Caller	
x18-27	s2-11	Callee	
x28-31	t3-t6	Caller	
f0-7	ft0-7	Caller	
f8-9	fs0-1	Callee	
f10-17	fa0-7	Caller	
f12-17	fa2-7	Caller	
f18-27	fs2-11	Callee	
f28-31	ft8-11	Caller	
zero			Hardwired zero
ra			Return address
sp			Stack pointer
gp			Global pointer
tp			Thread pointer
t0-0,ft0-7			Temporaries
s0-11,fs0-11			Saved registers
a0-7,fa0-7			Function args

RISC-V calling convention and five optional extensions: 8 RV32M; 11 RV32A; 34 floating-point instructions each for 32- and 64-bit data (RV32F, RV32D); and 53 RV32V. Using regex notation, {} means set, so FADD.{F|D} is both FADD.F and FADD.D. RV32{F|D} adds registers f0-f31, whose width matches the widest precision, and a floating-point control and status register fcsr. RV32V adds vector registers v0-v31, vector predicate registers vp0-vp7, and vector length register vl. RV64 adds a few instructions: RVM gets 4, RVA 11, RVF 6, RVD 6, and RVV 0.

The RISC-V Reader:
An Open Architecture Atlas
First Edition, 1.0.0

David Patterson and Andrew Waterman

November 7, 2017

Book version: 1.0.0

The cover background is a photo of the ***Mona Lisa***. It is a portrait of Lisa Gherardini, painted between 1503 and 1506, by the Leonardo da Vinci. The King of France bought it from Leonardo in about 1530, and it has been on display at the Louvre Museum in Paris since 1797. The Mona Lisa is considered the best known work of art in the world. Mona Lisa represents elegance, which we believe is a feature of RISC-V.

The book was prepared with LaTeX. The necessary Makefiles, style files, and most of the scripts are available under the BSD License at `github.com/armandofox/latex2ebook`.

Arthur Klepchukov designed the covers and graphics for all versions.

Publisher's Cataloging-in-Publication

Names: Patterson, David A. | Waterman, Andrew, 1986-
Title: The RISC-V reader: an open architecture atlas / David Patterson and Andrew Waterman.
Description: First edition. | [Berkeley, California] : Strawberry Canyon LLC, 2017. |
    Includes bibliographical references and index.
Identifiers: ISBN 978-0-9992491-1-6
Subjects: LCSH: Computer architecture. | RISC microprocessors. |
    Assembly languages (Electronic computers)
Classification: LCC QA76.9.A73 P38 2017 | DDC 004.22- -dc23

Dedication

David Patterson dedicates this book to his parents:

—To my father David, from whom I inherited inventiveness, athleticism, and the courage to fight for what is right; and

—To my mother Lucie, from whom I inherited intelligence, optimism, and my temperament.

Thank you for being such great role models, which taught me what it means to be a good spouse, parent, and grandparent.

Andrew Waterman dedicates this book to his parents, John and Elizabeth, who have been enormously supportive, even while thousands of miles away.

About the Authors

David Patterson retired after 40 years as a Professor of Computer Science at UC Berkeley in 2016, and then joined Google Brain as a distinguished engineer. He also serves as Vice-Chair of the Board of Directors of the RISC-V Foundation. In the past, he was named Chair of Berkeley's Computer Science Division and was elected to be Chair of the Computing Research Association and President of the Association for Computing Machinery. In the 1980s, he led four generations of Reduced Instruction Set Computer (RISC) projects, which inspired Berkeley's latest RISC to be named "RISC Five." Along with Andrew Waterman, he was one of the four architects of RISC-V. Beyond RISC, his best-known research projects are Redundant Arrays of Inexpensive Disks (RAID) and Networks of Workstations (NOW). This research led to many papers, 7 books, and more than 35 honors, including election to the National Academy of Engineering, the National Academy of Sciences, and the Silicon Valley Engineering Hall of Fame as well as being named a Fellow of the Computer History Museum, ACM, IEEE, and both AAAS organizations. His teaching awards include the Distinguished Teaching Award (UC Berkeley), the Karlstrom Outstanding Educator Award (ACM), the Mulligan Education Medal (IEEE), and the Undergraduate Teaching Award (IEEE). He also won Textbook Excellence Awards ("Texty") from the Text and Academic Authors Association for a computer architecture book and for a software engineering book. He received all his degrees from UCLA, which awarded him an Outstanding Engineering Academic Alumni Award. He grew up in Southern California, and for fun he plays soccer and rides bikes with his sons and walks on the beach with his wife. Originally high-school sweethearts, they celebrated their 50th wedding anniversary a few days after the Beta edition was published.

Andrew Waterman serves as SiFive's Chief Engineer and co-founder. SiFive was founded by the creators of the RISC-V architecture to provide low-cost custom chips based on RISC-V. Andrew received his PhD in Computer Science from UC Berkeley, where, weary of the vagaries of existing instruction set architectures, he co-designed the RISC-V ISA and the first RISC-V microprocessors. He is one of the main contributors to the open-source RISC-V-based Rocket chip generator, the Chisel hardware construction language, and the RISC-V ports of the Linux operating system kernel and the GNU C Compiler and C Library. He also has an MS from UC Berkeley, which was the basis of the RVC extension for RISC-V, and a BSE from Duke University.

Quick Contents

Contents

List of Figures

Preface

Welcome!

RISC-V has been a phenomenon, rapidly growing in popularity since its introduction in 2011. We thought a slim programmer's guide would help fuel its ride and encourage newcomers to understand why it is an attractive instruction set and see how it differs from conventional instruction set architectures (ISA) of the past.

Books for other ISAs inspired us, although we hoped that the simplicity of RISC-V would mean writing much less than the 500+ pages of fine books such as *See MIPS Run*. At one-third the overall length, at least by that measure we've succeeded. In fact, the ten chapters that introduce each component of the modular RISC-V instruction set take just 100 pages—despite averaging nearly one figure per page (75 total)—which makes for quick reading.

After explaining the principles of instruction set design, we show how the RISC-V architects learned from the instruction sets of the past 40 years to borrow their good ideas and avoid their mistakes. ISAs are judged as much by what is omitted as by what is included.

We then introduce each component of this modular architecture in a sequence of chapters. Every chapter has a program in RISC-V assembly language that demonstrates use of the instructions introduced in that chapter, which makes it easier for the assembly language programmer to learn RISC-V code. We also often show equivalent programs in ARM, MIPS, and x86 that highlight the simplicity and cost-energy-performance benefits of RISC-V.

To make the book more fun to read, we include almost 50 sidebars in the page margins with what we hope are interesting commentaries about the text. We also include about 75 images in the margins to emphasize examples of good ISA design. (Our margins are well-used!) Finally, for the dedicated reader, we add roughly 25 elaborations throughout the text. You can delve into these optional sections if you are interested in a topic. These sections aren't required to understand the other material in the book, so feel free to skip them if they don't catch your interest. For computer architecture buffs, we cite 25 papers and books that may broaden your horizons. We learned a lot by reading them in order to write this book!

Why So Many Quotes?

We think quotes also make the book more fun to read, so we're sprinkled 25 of them throughout the text. They likewise are an efficient mechanism to pass along wisdom from elders to novices, and help set cultural standards for good ISA design. We want readers to pick up a bit of history of the field too, which is why we feature quotes from famous computer scientists and engineers throughout the text.

Introduction and Reference

We intend this slim book to work as both an introduction and a reference to RISC-V for students and embedded systems programmers interested in writing RISC-V code. This book assumes readers have seen at least one instruction set beforehand. If not, you might want to browse our related introductory architecture book based on RISC-V: *Computer Organization and Design RISC-V Edition: The Hardware Software Interface*.

The compact references in this book include:

- **Reference Card** – This one page (two sides) condensed description of RISC-V covers both RV32GCV and RV64GCV, which includes the base and all defined extensions: RVI, RVM, RVA, RVF, RVD, RVC, and even RVV, even though it is still under development.

- **Instruction Diagrams** – These half-page graphical descriptions of each instruction extension, which are the first figures of the chapters, list the full names of all RISC-V instructions in a format that let's you easily see the variations of each instruction. See Figures 2.1, 4.1, 5.1, 6.1, 7.1, 8.1, 9.1, 9.2, 9.3, and 9.4.

- **Opcode Maps** – These tables show the instruction layout, opcodes, format type, and instruction mnemonic for each instruction extension in a fraction of a page. See Figures 2.3, 3.3, 3.4, 4.2, 5.2, 5.3, 6.2, 7.6, 7.5, 7.7, 9.5, and 10.1. (The instruction diagrams and opcode maps inspired the use of the word atlas in the book's subtitle.)

- **Instruction Glossary** – Appendix A is a thorough description of every RISC-V instruction and pseudoinstruction.[1] It includes everything: the operation name and operands, an English description, a register-transfer language definition, which RISC-V extension it is in, the full name of the instruction, the instruction format, a diagram of the instruction showing the opcodes, and references to compact versions of the instruction. Amazingly, this all fits into less than 50 pages.

- **Instruction Translator** – Appendix B helps experienced assembly language programmers by providing tables that show the ARM-32 or x86-32 instructions that are equivalent to RV32I instructions. It also lists the outputs of the C compiler for a simple tree-traversal program for these three architectures and describes the surprisingly small differences between them. It concludes with advice on how to translate code from the older architectures into RISC-V, which is easier than one might think.

- **Index** – It helps you find the page that describes the instruction explanation, definition, or diagram either by the full name or by mnemonic. It is organized like a dictionary.

---

[1] The committee defining RV32V did not complete their work in time for the Beta edition, so we omit those instructions from Appendix A. Chapter 8 is our best guess of what RV32V will be, although it is likely to change a little.

Errata and Supplementary Content

We intend to collect the Errata together and release updates a few times a year. The book's website shows the latest version of the book and a brief description of the changes since the previous version. Previous errata can be reviewed, and new ones reported, on the book's website (www.riscvbook.com). We apologize in advance for the problems you find in this edition, and look forward to your feedback on how to improve this material.

History of this Book

At the Sixth RISC-V Workshop held May 8–11, 2017 in Shanghai, we saw the need for such a book. We started a few weeks later. Given Patterson's much greater experience in writing books, the plan was for him to write most chapters. Both of us collaborated on the organization and were first reviewers for each other's chapters. Patterson authored Chapters 1, 2, 3, 4, 5, 6, 7, 8, 9, 11, the Reference Card, and this Preface, while Waterman wrote 10, Appendix A—the largest section of the book—Appendix B, and coded all the programs in the book. Waterman also maintained the LaTeX pipeline from Armando Fox that let us produce the book.

We offered a Beta edition of the textbook for 800 UC Berkeley students in the Fall semester 2017. Readers only found a few typos and LaTeX bugs, which we fixed for the first edition. We also improved the margin icons to make them easier to remember and revised a few figures that didn't look as good on the printed page as we hoped.

More significantly, the first edition expanded Chapter 10 to include 60+ Control and Status Registers and added Appendix B to help programmers interested in converting assembly language programs from the older ISAs into RISC-V.

The first edition was be published in time to be available at the Seventh RISC Workshop in Silicon Valley from November 28–30, 2017.

RISC-V was a byproduct of a Berkeley research project[1] that was developing technology to make it easier to build parallel hardware and software.

Acknowledgments

We wish to thank Armando Fox for use of his LaTeX pipeline and advice on navigating the world of self publishing.

Our deepest thanks go to the people who read early drafts of the book and offered helpful suggestions: Krste Asanović, Nikhil Athreya, C. Gordon Bell, Stuart Hoad, David Kanter, John Mashey, Ivan Sutherland, Ted Speers, Michael Taylor, and Megan Wachs.

Finally, we thank the many UC Berkeley students for their debugging help and their continuing interest in this material!

David Patterson and Andrew Waterman
November 16, 2017
Berkeley, California

1 Why RISC-V?

Leonardo da Vinci
(1452-1519) was a Renaissance architect, engineer, sculptor, and painter of the Mona Lisa.

Simplicity is the ultimate sophistication.

—Leonardo da Vinci

1.1 Introduction

The goal for RISC-V ("RISC five") is to become a universal *instruction set architecture* (*ISA*):

- It should suit all sizes of processors, from the tiniest embedded controller to the fastest high-performance computer.

- It should work well with a wide variety of popular software stacks and programming languages.

- It should accommodate all implementation technologies: Field-Programmable Gate Arrays (FPGAs), Application-Specific Integrated Circuits (ASICs), full-custom chips, and even future device technologies.

- It should be efficient for all microarchitecture styles: microcoded or hardwired control; in-order, decoupled, or out-of-order pipelines; single or superscalar instruction issue; and so on.

- It should support extensive specialization to act as a base for customized accelerators, which rise in importance as Moore's Law fades.

- It should be stable, in that the base ISA should not change. More importantly, the ISA cannot be discontinued, as has happened in the past to proprietary ISAs such as the AMD Am29000, the Digital Alpha, the Digital VAX, the Hewlett Packard PA-RISC, the Intel i860, the Intel i960, the Motorola 88000, and the Zilog Z8000.

We add sidebars in the margins to offer hopefully interesting commentary. For example, RISC-V was originally developed for internal use in UC Berkeley research and courses. It became open because outsiders started using it on their own. The RISC-V architects learned about the external interest when they started receiving complaints about ISA changes in their coursework, which was on the web. Only after the architects understood the need did they try to make it an open ISA standard.

RISC-V is unusual not only because it is a recent ISA—born this decade when most alternatives date from the 1970s or 1980s—but also because it is an *open* ISA. Unlike practically all prior architectures, its future is free from the fate or the whims of any single corporation, which has doomed many ISAs in the past. It belongs instead to an open, non-profit foundation. The goal of the RISC-V Foundation is to maintain the stability of RISC-V, evolve it slowly and carefully, solely for technical reasons, and try to make it as popular for hardware as Linux is for operating systems. As a sign of its vitality, Figure 1.1 lists the largest corporate members of the RISC-V Foundation.

>$50B		>$5B, <$50B		>$0.5B, <$5B	
Google	USA	BAE Systems	UK	AMD	USA
Huawei	China	MediaTek	Taiwan	Andes Technology	China
IBM	USA	Micron Tech.	USA	C-SKY Microsystems	China
Microsoft	USA	Nvidia	USA	Integrated Device Tech.	USA
Samsung	Korea	NXP Semi.	Netherlands	Mellanox Technology	Israel
		Qualcomm	USA	Microsemi Corp.	USA
		Western Digital	USA		

Figure 1.1: The corporate members of the RISC-V Foundation as of the Sixth RISC-V Workshop in May 2017 ranked by annual sales. The left column companies all exceed $US 50B in annual sales, the middle column companies sell less than $US 50B but more than $US 5B, and the sales of those in the right column are less than $US 5B but more than $US 0.5B. The foundation includes another 25 smaller companies, 5 startup companies (Antmicro Ltd, Blockstream, Esperanto Technologies, Greenwaves Technologies, and SiFive), 4 nonprofit organizations (CSEM, Draper Laboratory, ICT, and lowRISC), and 6 universities (ETH Zurich, IIT Madras, National University of Defense Technology, Princeton, and UC Berkeley). Most of the 60 organizations have their headquarters outside the US. To learn more, see www.riscv.org.

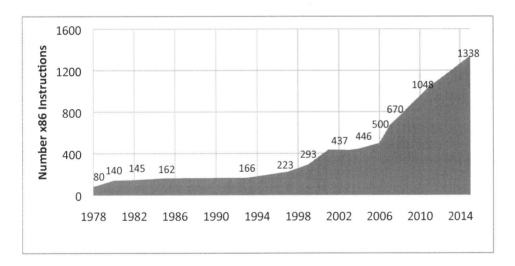

Figure 1.2: Growth of x86 instruction set over its lifetime. x86 started with 80 instructions in 1978. It grew 16X to 1338 instructions by 2015, and it's still growing. Amazingly, this graph is conservative. An Intel blog puts the count at 3600 instructions in 2015 [Rodgers and Uhlig 2017], which would raise the x86 rate to one new instruction *every four days* between 1978 and 2015. We count assembly language instructions, and they presumably count machine language instructions. As Chapter 8 explains, a large part of the growth is because the x86 ISA relies on SIMD instructions for data level parallelism.

```
The AL register is the default source and destination.
If the low 4-bits of AL register are > 9,
   or the auxiliary carry flag AF = 1,
Then
    Add 6 to low 4-bits of AL and discard overflow
    Increment the high byte of AL
    Carry flag CF = 1
    Auxiliary carry flag AF = 1
Else
    CF = AF = 0
Upper 4-bits of AL = 0
```

Figure 1.3: Description of the x86-32 *ASCII Adjust after Addition* (aaa) instruction. It performs computer arithmetic in Binary Coded Decimal (BCD), which has fallen into the dustbin of information technology history. The x86 also has three related instructions for subtraction (aas), multiplication (aam), and division (aad). As each is a one-byte instruction, they collectively occupy 1.6% (4/256) of the precious opcode space.

1.2 Modular vs. Incremental ISAs

> *Intel was betting its future on a high-end microprocessor, but that was still years away. To counter Zilog, Intel developed a stop-gap processor and called it the 8086. It was intended to be short-lived and not have any successors, but that's not how things turned out. The high-end processor ended up being late to market, and when it did come out, it was too slow. So the 8086 architecture lived on—it evolved into a 32-bit processor and eventually into a 64-bit one. The names kept changing (80186, 80286, i386, i486, Pentium), but the underlying instruction set remained intact.*
>
> —Stephen P. Morse, architect of the 8086 [Morse 2017]

The conventional approach to computer architecture is *incremental* ISAs, where new processors must implement not only new ISA extensions but also all extensions of the past. The purpose is to maintain *backwards binary-compatibility* so that binary versions of decades-old programs can still run correctly on the latest processor. This requirement, when combined with the marketing appeal of announcing new instructions with a new generation of processors, has led to ISAs that grow substantially in size with age. For example, Figure 1.2 shows the growth in the number of instructions for a dominant ISA today: the 80x86. It dates back to 1978, yet it has added about *three instructions per month* over its long lifetime.

This convention means that every implementation of the x86-32 (the name we use for the 32-bit address version of x86) must implement the mistakes of past extensions, even when they no longer make sense. For example, Figure 1.3 describes the ASCII Adjust after Addition (aaa) instruction of the x86, which has long outlived its usefulness.

As an analogy, suppose a restaurant serves only a fixed-price meal, which starts out as a small dinner of just a hamburger and a milkshake. Over time, it adds fries, and then an ice cream sundae, followed by salad, pie, wine, vegetarian pasta, steak, beer, ad infinitum until it becomes a gigantic banquet. It may make little sense in total, but diners can find whatever they've ever eaten in a past meal at that restaurant. The bad news is that diners must pay the rising cost of the expanding banquet for each dinner.

Beyond being recent and open, RISC-V is unusual since, unlike almost all prior ISAs, it is *modular*. At the core is a base ISA, called *RV32I*, which runs a full software stack. RV32I is frozen and will never change, which gives compiler writers, operating system developers, and assembly language programmers a stable target. The modularity comes from optional standard extensions that hardware can include or not depending on the needs of the application. This modularity enables very small and low energy implementations of RISC-V, which can be critical for embedded applications. By informing the RISC-V compiler what extensions are included, it can generate the best code for that hardware. The convention is to append the extension letters to the name to indicate which are included. For example, RV32IMFD adds the multiply (RV32M), single-precision floating point (RV32F), and double-precision floating point extensions (RV32D) to the mandatory base instructions (RV32I).

If software uses an omitted RISC-V instruction from an optional extension, the hardware traps and executes the desired function in software as part of a standard library.

Returning to our analogy, RISC-V offers a menu instead of a buffet; the chef need cook only what the customers want—not a feast for every meal—and the customers pay only for what they order. RISC-V has no need to add instructions simply for the marketing sizzle. The RISC-V Foundation decides when to add a new option to the menu, and they will do so only for solid technical reasons after an extended open discussion by a committee of hardware and software experts. Even when new choices appear on the menu, they remain optional and not a new requirement for all future implementations, like incremental ISAs.

1.3 ISA Design 101

Before introducing the RISC-V ISA, it will be helpful to understand the underlying principles and trade-offs that a computer architect must make while designing an ISA. Below is a list of the seven measures, along with icons we'll put in page margins to highlight instances when RISC-V addresses them in the following chapters. (The back cover of the print book has a legend for the icons.)

- cost (US dollar coin icon)

- simplicity (wheel)

- performance (speedometer)

- isolation of architecture from implementation (detached halves of a circle)

- room for growth (accordion)

- program size (opposing arrows compressing line)

- ease of programming / compiling / linking ("as easy as ABC").

Cost

Simplicity

Performance

Isolation of Arch from Impl

Room for Growth

Code Size

Programmability

To illustrate what we mean, in this section we'll show some choices from older ISAs that look unwise in retrospect and where RISC-V often made much better decisions.

Cost. Processors are implemented as integrated circuits, commonly called *chips* or *dies*. They are called dies because they start life as a piece of a single round wafer, which is *diced* into many individual pieces. Figure 1.4 shows a wafer of RISC-V processors. The cost is very sensitive to the area of the die:

$$cost \approx f(die\ area^2)$$

Obviously, the smaller the die, the more dies per wafer, and most of the cost of the die is the processed wafer itself. Less obvious is that the smaller the die, the higher the *yield*, the

Figure 1.4: An 8-inch diameter wafer of RISC-V dies designed by SiFive. It has two types of RISC-V dies using an older, larger processing line. An FE310 die is 2.65 mm×2.72 mm and a SiFive test die that is 2.89 mm×2.72 mm. The wafer contains 1846 of the former and 1866 of the latter, totaling 3712 chips.

fraction of manufactured dies that work. The reason is that the silicon manufacturing will result in small flaws scattered about the wafer, so the smaller the die, the lower the fraction that will be flawed.

An architect wants to keep the ISA simple to shrink the size of processors that implement it. As we shall see in the following chapters, the RISC-V ISA is much simpler ISA than the ARM-32 ISA. As a concrete example of the impact of simplicity, let's compare a RISC-V Rocket processor to an ARM-32 Cortex-A5 processor in the same technology (TSMC40GPLUS) using the same-sized caches (16 KiB). The RISC-V die is 0.27 mm^2 versus 0.53 mm^2 for ARM-32. Around twice the area, the ARM-32 Cortex-A5 die costs approximately 4X (2^2) as much as RISC-V Rocket die. Even a 10% smaller die reduces cost by a factor of 1.2 (1.1^2).

Simplicity.   Given the cost sensitivity to complexity, architects want a simple ISA to reduce die area. Simplicity also reduces chip design time and verification time, which can be much of the cost of development of the chip. These costs must be added to the cost of the chip, with this overhead dependent on the number of chips shipped. Simplicity also reduces the cost of documentation and the difficulty of getting customers to understand how to use the ISA.

Below is a glaring example of ISA complexity from ARM-32:

```
ldmiaeq SP!, {R4-R7, PC}
```

The instruction stands for LoaD Multiple, Increment-Address, on EQual. It performs 5 data loads and writes to 6 registers but executes only if the EQ condition code is set. Moreover, it writes a result to the PC, so it is also performing a conditional branch. Quite a handful!

Ironically, simple instructions are much more likely to be used than complex ones. For example, x86-32 includes an `enter` instruction, which was intended to be the first instruction executed on entering a procedure to create a stack frame for it (see Chapter 3). Most compilers instead use only these two simple x86-32 instructions:

```
push ebp      # Push the frame pointer onto the stack
mov  ebp, esp # Copy the stack pointer to the frame pointer
```

Performance.   Except for the tiny chips for embedded applications, architects are typically concerned about performance as well as cost. Performance can be factored into three terms:

$$\frac{instructions}{program} \times \frac{average\ clock\ cycles}{instruction} \times \frac{time}{clock\ cycle} = \frac{time}{program}$$

Even if a simple ISA might execute more instructions per program than a complex ISA, it can more than make up for that by having a faster clock cycle or average fewer clock cycles per instruction (CPI).

For example, for the CoreMark benchmark [Gal-On and Levy 2012] (100,000 iterations), the performance on the ARM-32 Cortex-A9 is

$$\frac{32.27\ B\ instructions}{program} \times \frac{0.79\ clock\ cycles}{instruction} \times \frac{0.71\ ns}{clock\ cycle} = \frac{18.15\ secs}{program}$$

For the BOOM implementation of RISC-V, the equation is

$$\frac{29.51\ B\ instructions}{program} \times \frac{0.72\ clock\ cycles}{instruction} \times \frac{0.67\ ns}{clock\ cycle} = \frac{14.26\ secs}{program}$$

High-end processors can gain performance by combining simple instructions together without burdening all lower-end implementations with a larger, more complicated ISA. This technique is called *macrofusion*, as it fuses "macro" instructions together.

Simplicity

A simple processor can be helpful for embedded applications since it is easier to predict execution time. Assembly-language programmers of microcontrollers often want to maintain exact timing, so they rely on code taking a predictable number of clock cycles that they can count by hand.

The last factor is the inverse of the clock rate, so a 1 GHz clock rate means the time per clock cycle is 1 ns ($1/10^9$).

The average number of clock cycles can be less than 1 because the A9 and BOOM [Celio et al. 2015] are so-called *superscalar* processors, which execute more than one instruction per clock cycle.

The ARM processor didn't execute fewer instructions than RISC-V in this case. As we shall see, the simple instructions are also the most popular instructions, so ISA simplicity can win in all metrics. For this program, the RISC-V processor gains nearly 10% in each of the three factors, which results in a performance advantage of almost 30%. If a simpler ISA also results in a smaller chip, its cost-performance will be excellent.

Isolation of Architecture from Implementation. The original distinction between *architecture* and *implementation*, which goes back to the 1960s, is that architecture is what a machine language programmer needs to know to write a correct program, but not the performance of that program. The temptation for an architect is to include instructions in an ISA that help performance or cost of one implementation at a particular time, but burden different or future implementations.

Isolation of Arch from Impl

> **Pipelined processors today anticipate branch outcomes** using hardware predictors, which can exceed 90% accuracy and work with any pipeline length. They only need a mechanism to flush and restart the pipeline when they mispredict.

For the MIPS-32 ISA, the regrettable example was the *delayed branch*. Conditional branches cause problems in pipelined execution because the processor wants to have the next instruction to execute already in the pipeline, but it can't decide whether it wants the next sequential one (if the branch isn't taken) or the one at the branch target address (if it is taken). For their first microprocessor with a 5-stage pipeline, this indecision could have caused a one clock-cycle stall of the pipeline. MIPS-32 solved this problem by redefining branch to occur in the instruction *after* the next one. Thus, the following instruction is *always* executed. The job of the programmer or compiler writer was to put something useful into the *delay slot*.

Alas, this "solution" didn't help later MIPS-32 processors with many more pipeline stages (hence many more instructions fetched before the branch outcome is computed), but it made life harder for MIPS-32 programmers, compiler writers, and processor designers ever after, since incremental ISAs demand backwards compatibility (see Section 1.2). In addition, it makes the MIPS-32 code much harder to understand (see Figure 2.10 on page 29).

While architects shouldn't put features that *help* just one implementation at a point in time, they also shouldn't put in features that *hinder* some implementations. For example, ARM-32 and some other ISAs have a Load Multiple instruction, as mentioned on the previous page. These instructions can improve performance of single-instruction issue pipelined designs, but hurt multiple-instruction issue pipelines. The reason is that the straightforward implementation precludes scheduling the individual loads of a Load Multiple in parallel with other instructions, reducing instruction throughput of such processors.

Room for Growth

Room for Growth.   With ending of Moore's Law, the only path forward for major improvements in cost-performance is to add custom instructions for specific domains, such as deep learning, augmented reality, combinatorial optimization, graphics, and so. That means it's important today for an ISA to reserve opcode space for future enhancements.

In the 1970s and 1980s, when Moore's Law was in full force, there was little thought of saving opcode space for future accelerators. Architects instead valued larger address and immediate fields to reduce the number of instructions executed per program, the first factor in the performance equation on the prior page.

> **The ARM-32 instruction** `ldmiaeq` **mentioned above is even more complicated**, since when it branches it can also change instruction set modes between ARM-32 and Thumb/Thumb-2.

An example of the impact of paucity of opcode space was when the architects of ARM-32 later tried to reduce code size by adding 16-bit length instructions to the formerly uniform 32-bit length ISA. There was simply no room left. Thus, the only solution was to create a new ISA first with 16-bit instructions (Thumb) and later a new ISA with both 16-bit and 32-bit instructions (Thumb-2) using a mode bit to switch between ARM ISAs. To change modes, the programmer or compiler branches to a byte address with a 1 in the least-significant bit, which worked because 16-bit and 32-bit instructions should have 0 in that bit.

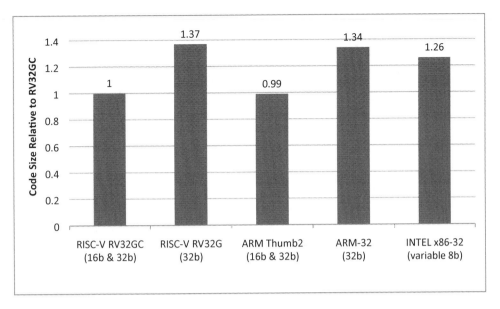

Figure 1.5: Relative program sizes for RV32G, ARM-32, x86-32, RV32C, and Thumb-2. The last two ISAs are aimed at small code size. The programs were the SPEC CPU2006 benchmarks using the GCC compilers. The small size advantage of Thumb-2 over RV32C is due to the code size savings of Load and Store Multiple on procedure entry. RV32C excludes them to maintain the one-to-one mapping to instructions of RV32G, which omits Load and Store Multiple to reduce implementation complexity for high-end processors (see below). Chapter 7 explains RV32C. RV32G indicates a popular combination of RISC-V extensions (RV32M, RV32F, RV32D, and RV32A), properly called RV32IMAFD. [Waterman 2016]

Program Size. The smaller the program, the smaller the area on a chip needed for the program memory, which can be a significant cost for embedded devices. Indeed, that issue inspired ARM architects to retroactively add shorter instructions in the Thumb and Thumb-2 ISAs. Smaller programs also lead to fewer misses in instruction caches, which saves power since off-chip DRAM accesses use much more energy than on-chip SRAM accesses, and improves performance as well. Small code size can be one of the goals of ISA architects.

The x86-32 ISA has instructions as short as 1 byte and as long as 15 bytes. One would expect that the byte-variable length instructions of the x86 should certainly lead to smaller programs than ISAs limited to 32-bit length instructions, like ARM-32 and RISC-V. Logically, 8-bit variable length instructions should also be smaller than ISAs that offer only 16-bit and 32-bit instructions, like Thumb-2 and RISC-V using the RV32C extension (see Chapter 7). Figure 1.5 shows that, while ARM-32 and RISC-V code is 6% to 9% larger than code for x86-32 when all instructions are 32 bits long, surprisingly x86-32 is 26% *larger* than the compressed versions (RV32C and Thumb-2) that offer both 16-bit and 32-bit instructions.

While a new ISA using 8-bit variable instructions would likely lead to smaller code than RV32C and Thumb-2, the architects of the first x86 in the 1970s had different concerns. Moreover, given the requirement of backwards binary-compatibility of an incremental ISA (Section 1.2), the hundreds of new x86-32 instructions are longer than one might expect, since they bear the burden of a one- or two-byte prefix to squeeze them into the limited free opcode space of the original x86.

Code Size

One example 15-byte x86-32 instruction is `lock add dword ptr ds:[esi+ecx*4 +0x12345678], 0xefcdab89`. It assembles into (in hexadecimal): 67 66 f0 3e 81 84 8e 78 56 34 12 89 ab cd ef. The last 8 bytes are 2 addresses and the first 7 bytes specify atomic memory operation, the add operation, 32-bit data, the data segment register, the 2 address registers, and scaled indexed addressing mode. An example 1-byte instruction is `inc eax` that assembles into 40.

Programmability

Ease of programming, compiling, and linking.   Since data in a register is so much faster to access than data in memory, it is critical for compilers to do a good job at register allocation. That task is much easier when there are many registers rather than fewer. In that light, ARM-32 has 16 registers and x86-32 has only 8. Most modern ISAs, including RISC-V, have a relatively generous 32 integer registers. More registers surely make life easier for compilers and assembly language programmers.

Another issue for compilers and assembly language programmers is figuring out the speed of a code sequence. As we shall see, RISC-V instructions are typically at most one clock cycle per instruction (ignoring cache misses), while as we saw earlier both ARM-32 and x86-32 have instructions that take many clock cycles even when everything fits in the cache. Moreover, unlike ARM-32 and RISC-V, x86-32 arithmetic instructions can have operands in memory instead of requiring all operands to be in registers. Complex instructions and operands in memory make it difficult for processor designers to deliver performance predictability.

It's useful for an ISA to support *position independent code* (*PIC*), because it supports dynamic linking (see Section 3.5), since shared library code can reside at different addresses in different programs. PC-relative branches and data addressing are a boon to PIC. While nearly all ISAs provide PC-relative branches, x86-32 and MIPS-32 omit PC-relative data addressing.

■ *Elaboration: ARM-32, MIPS-32, and x86-32*

Elaborations are optional sections that readers can delve into if they are interested in a topic, but you don't need to read them to understand the rest of the book. For example, our ISA names aren't the official ones. The 32-bit-address ARM ISA has many versions, with the first in 1986 and the latest called ARMv7 in 2005. ARM-32 generally refers to the ARMv7 ISA. MIPS also had many 32-bit versions, but we're referring to the original, called MIPS I. ("MIPS32" is a different, later ISA than what we call MIPS-32.) Intel's first 16-bit address architecture was the 8086 in 1978, which the 80386 ISA expanded to 32-bit addresses in 1985. Our x86-32 notation generally refers to the IA-32, the 32-bit-address version of its x86 ISA. Given the myriad variants of these ISAs, we find our nonstandard terminology least confusing.

1.4   An Overview of this Book

This book assumes you have seen other instruction sets before RISC-V. If not, look at our related introductory architecture book based on RISC-V [Patterson and Hennessy 2017].

Chapter 2 introduces RV32I, the frozen base integer instructions that are the heart of RISC-V. Chapter 3 explains the remaining RISC-V assembly language beyond that introduced in Chapter 2, including calling conventions and some clever tricks for linking. Assembly language includes all of the proper RISC-V instructions plus some useful instructions that are outside RISC-V. These *pseudoinstructions*, which are clever variations of real instructions, make it easier to write assembly language programs without having to complicate the ISA.

The next three chapters explain the standard RISC-V extensions that, when added to RV32I, we collectively call RV32G (G is for general):

- Chapter 4: Multiply and Divide (RV32M)

- Chapter 5: Floating Point (RV32F and RV32D)

- Chapter 6: Atomic (RV32A)

The RISC-V "reference card" on pages 3 and 4 is a handy summary of *all* RISC-V instructions in this book: RV32G, RV64G, and RV32/64V.

Chapter 7 describes the optional compressed extension RV32C, an excellent example of the elegance of RISC-V. By restricting the 16-bit instructions to be short versions of existing 32-bit RV32G instructions, they are almost free. The assembler can pick the instruction size, allowing the assembly language programmer and the compiler to be oblivious to RV32C. The hardware decoder to translate 16-bit RV32C instructions into 32-bit RV32G instructions needs just 400 gates, which is a few percent of even the simplest implementation of RISC-V.

Chapter 8 introduces RV32V, the vector extension. Vector instructions are another example of ISA elegance as compared to the numerous, brute-force *Single Instruction Multiple Data (SIMD)* instructions of ARM-32, MIPS-32, and x86-32. Indeed, hundreds of the instructions added to x86-32 in Figure 1.2 were SIMD, and hundreds more are coming. RV32V is even simpler than most vector ISAs, as it associates the data type and length with the vector registers instead of embedding them in the opcodes. RV32V may be the most compelling reason for switching from a conventional SIMD-based ISA to RISC-V.

Chapter 9 shows the 64-bit address version of RISC-V, RV64G. As the chapter explains, the RISC-V architects needed only to widen the registers and add a few word, doubleword, or long versions of RV32G instructions to extend the address from 32 to 64 bits.

Chapter 10 explains the system instructions, showing how RISC-V handles paging and the Machine, User, and Supervisor privilege modes.

The last chapter gives a quick description of the remaining extensions that are currently under consideration by the RISC-V Foundation.

Next comes the largest section of the book, Appendix A, an instruction set summary in alphabetical order. It defines the full RISC-V ISA with all extensions mentioned above and all pseudoinstructions in about 50 pages, a testimony to the simplicity of RISC-V.

Appendix B shows common assembly language operations and what instructions they correspond to in RV32I, ARM-32, and x86-32. Those three figures are followed by small C program and the output of the compiler for three ISAs. The appendix servers two purposes. For readers already familiar with ARM-32 or x86-32 ISAs, it as another way to learn RISC-V by mapping it the more ISA they know well. The second purpose is to aid programmers who are translating existing assembly language programs in these older ISAs into RISC-V.

We end the book with an index.

1.5 Concluding Remarks

> *It is easy to see by formal-logical methods that there exist certain [instruction sets] that are in abstract adequate to control and cause the execution of any sequence of operations ... The really decisive considerations from the present point of view, in selecting an [instruction set], are more of a practical nature: simplicity of the equipment demanded by the [instruction set], and the clarity of its application to the actually important problems together with the speed of its handling of those problems.*
>
> —[von Neumann et al. 1947, 1947]

The reference card is also called the *green card* because of the shade of the background color of the one-page cardboard summary of ISAs from the 1960s. We kept the background white for legibility instead of green for historical accuracy.

Simplicity

A previous version of John von Neumann's well-written report was so influential that this style of computer is commonly called a *von Neumann architecture*, although this report was based on the work of others. It was written three years before the first stored program computer was operational!

ISA	Pages	Words	Hours to read	Weeks to read
RISC-V	236	76,702	6	0.2
ARM-32	2736	895,032	79	1.9
x86-32	2198	2,186,259	182	4.5

Figure 1.6: Number of pages and words of ISA manuals [Waterman and Asanović 2017a], [Waterman and Asanović 2017b], [Intel Corporation 2016], [ARM Ltd. 2014]. Hours and weeks to complete assumes reading at 200 words per minute for 40 hours a week. Based in part of Figure 1 of [Baumann 2017].

RISC-V is a recent, clean-slate, minimalist, and open ISA informed by mistakes of past ISAs. The goal of the RISC-V architects is for it to be effective for all computing devices, from the smallest to the fastest. Following von Neumann's 70-year-old advice, this ISA emphasizes simplicity to keep costs low while having plenty of registers and transparent instruction speed to help compilers and assembly language programmers map actually important problems to appropriate, quick code.

Simplicity

One indication of complexity is the size of the documentation.   Figure 1.6 shows the size of the instruction set manuals for RISC-V, ARM-32, and x86-32 measured in pages and words. If you read manuals as a full-time job—8 hours a day for 5 days a week—it would take half a month to make a single pass over the ARM-32 manual and a full month for the x86-32. At this level of intricacy, perhaps no single person fully understands ARM-32 or x86-32. Using this common-sense metric, RISC-V is $\frac{1}{12}$ complexity of the ARM-32 and $\frac{1}{10}$ to $\frac{1}{30}$ the complexity of x86-32. Indeed, the summary of RISC-V ISA including all extensions is only two pages (see the Reference Card).

This minimal, open ISA was unveiled in 2011 and is now backed by a foundation that will evolve it by adding optional extensions based strictly on technical justifications after a prolonged debate. The openness enables free, shared implementations of RISC-V, which lowers costs and the odds of unwanted malicious secrets being hidden in a processor.

However, hardware alone does not a system make. Software development costs likely dwarf hardware development costs, so while stable hardware is important, stable software is more so. It needs operating systems, boot-loaders, reference software, and popular software tools. The foundation offers stability for the overall ISA, and the frozen base means that the RV32I core that is the target for the software stack will never change. By its broad adoption and openness, RISC-V can challenge the dominance of the prevailing proprietary ISAs.

Elegance

Elegant is a word rarely applied to ISAs, but after reading this book, you may agree with us that it applies to RISC-V. We'll highlight features that we believe indicate elegance with a Mona Lisa icon in the margins.

1.6   To Learn More

ARM Ltd. ARM Architecture Reference Manual: ARMv7-A and ARMv7-R Edition, 2014. URL http://infocenter.arm.com/help/topic/com.arm.doc.ddi0406c/.

A. Baumann. Hardware is the new software. In *Proceedings of the 16th Workshop on Hot Topics in Operating Systems*, pages 132–137. ACM, 2017.

C. Celio, D. Patterson, and K. Asanovic. The Berkeley Out-of-Order Machine (BOOM): an industry-competitive, synthesizable, parameterized RISC-V processor. *Tech. Rep. UCB/EECS-2015–167, EECS Department, University of California, Berkeley*, 2015.

S. Gal-On and M. Levy. Exploring CoreMark - a benchmark maximizing simplicity and efficacy. *The Embedded Microprocessor Benchmark Consortium*, 2012.

Intel Corporation. *Intel 64 and IA-32 Architectures Software Developer's Manual, Volume 2: Instruction Set Reference*. September 2016.

S. P. Morse. The Intel 8086 chip and the future of microprocessor design. *Computer*, 50(4): 8–9, 2017.

D. A. Patterson and J. L. Hennessy. *Computer Organization and Design RISC-V Edition: The Hardware Software Interface*. Morgan Kaufmann, 2017.

S. Rodgers and R. Uhlig. X86: Approaching 40 and still going strong, 2017.

J. L. von Neumann, A. W. Burks, and H. H. Goldstine. Preliminary discussion of the logical design of an electronic computing instrument. *Report to the U.S. Army Ordnance Department*, 1947.

A. Waterman. *Design of the RISC-V Instruction Set Architecture*. PhD thesis, EECS Department, University of California, Berkeley, Jan 2016. URL http://www2.eecs.berkeley. edu/Pubs/TechRpts/2016/EECS-2016-1.html.

A. Waterman and K. Asanović, editors. *The RISC-V Instruction Set Manual Volume II: Privileged Architecture Version 1.10*. May 2017a. URL https://riscv.org/ specifications/privileged-isa/.

A. Waterman and K. Asanović, editors. *The RISC-V Instruction Set Manual, Volume I: User-Level ISA, Version 2.2*. May 2017b. URL https://riscv.org/specifications/.

Notes

[1] http://parlab.eecs.berkeley.edu

2   RV32I: RISC-V Base Integer ISA

Frances Elizabeth "Fran" Allen (1932-) was bestowed the Turing Award primarily for her work on optimizing compilers. The Turing Award is the greatest prize in Computer Science.

...the only way to realistically realize the performance goals and make them accessible to the user was to design the compiler and the computer at the same time. In this way features would not be put in the hardware which the software could not use ...

—Frances Elizabeth "Fran" Allen, 1981

2.1   Introduction

Figure 2.1 is a one-page graphical representation of the RV32I base instruction set. You can see the full RV32I instruction set by concatenating the underlined letters from left to right for each diagram. The set notation using { } lists the possible variations of the instruction, using either underlined letters or the underscore character _, which means no letter for this variation. For example

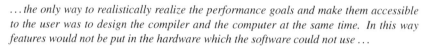

represents these four RV32I instructions: `slt`, `slti`, `sltu`, `sltiu`.

The goal of these diagrams, which will be the first figure of the following chapters, is give a quick, insightful overview of the instructions of a chapter.

2.2   RV32I Instruction formats

Simplicity

Cost

Performance

Figure 2.2 shows the six base instruction formats: R-type for register-register operations; I-type for short immediates and loads; S-type for stores; B-type for conditional branches; U-type for long immediates; and J-type for unconditional jumps. Figure 2.3 lists the opcodes of the RV32I instructions in Figure 2.1 using the formats of Figure 2.2.

Even the instruction formats demonstrate several examples where the simpler RISC-V ISA improves cost-performance. First, there are only six formats and all instructions are 32 bits long, which simplifies instruction decoding. ARM-32 and particularly x86-32 have numerous formats, which make decoding expensive in low-end implementations and a performance challenge for medium and high-end processor designs. Second, RISC-V instructions offer three register operands, rather than having one field shared for source and destination, as with x86-32. When an operation naturally has three distinct operands but the ISA provides

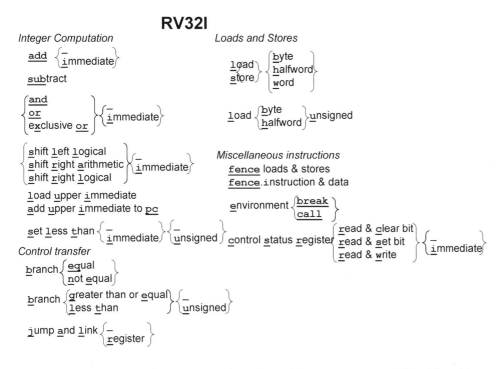

Figure 2.1: Diagram of the RV32I instructions. The underlined letters are concatenated from left to right to form RV32I instructions. The curly bracket notation { } means each vertical item in the set is a different variation of the instruction. The underscore _ within a set means that one option is simply the instruction name so far without a letter from this set. For example, the notation near the upper left-hand corner represents the following six instructions: and, or, xor, andi, ori, xori.

31	30	25	24	21	20	19	15	14	12	11	8	7	6	0	
funct7			rs2			rs1		funct3		rd			opcode		R-type
imm[11:0]						rs1		funct3		rd			opcode		I-type
imm[11:5]			rs2			rs1		funct3		imm[4:0]			opcode		S-type
imm[12]	imm[10:5]		rs2			rs1		funct3		imm[4:1]		imm[11]	opcode		B-type
imm[31:12]										rd			opcode		U-type
imm[20]	imm[10:1]			imm[11]		imm[19:12]				rd			opcode		J-type

Figure 2.2: RISC-V instruction formats. We label each immediate subfield with the bit position (imm[x]) in the immediate value being produced, rather than the bit position in the instruction's immediate field as is usually done. Chapter 10 explains how the control status register instructions use the I-type format slightly differently. (Figure 2.2 of Waterman and Asanović 2017 is the basis of this figure).

31 25	24 20	19 15	14 12	11 7	6 0		
imm[31:12]				rd	0110111	U	lui
imm[31:12]				rd	0010111	U	auipc
imm[20|10:1|11|19:12]				rd	1101111	J	jal
imm[11:0]		rs1	000	rd	1100111	I	jalr
imm[12|10:5]	rs2	rs1	000	imm[4:1|11]	1100011	B	beq
imm[12|10:5]	rs2	rs1	001	imm[4:1|11]	1100011	B	bne
imm[12|10:5]	rs2	rs1	100	imm[4:1|11]	1100011	B	blt
imm[12|10:5]	rs2	rs1	101	imm[4:1|11]	1100011	B	bge
imm[12|10:5]	rs2	rs1	110	imm[4:1|11]	1100011	B	bltu
imm[12|10:5]	rs2	rs1	111	imm[4:1|11]	1100011	B	bgeu
imm[11:0]		rs1	000	rd	0000011	I	lb
imm[11:0]		rs1	001	rd	0000011	I	lh
imm[11:0]		rs1	010	rd	0000011	I	lw
imm[11:0]		rs1	100	rd	0000011	I	lbu
imm[11:0]		rs1	101	rd	0000011	I	lhu
imm[11:5]	rs2	rs1	000	imm[4:0]	0100011	S	sb
imm[11:5]	rs2	rs1	001	imm[4:0]	0100011	S	sh
imm[11:5]	rs2	rs1	010	imm[4:0]	0100011	S	sw
imm[11:0]		rs1	000	rd	0010011	I	addi
imm[11:0]		rs1	010	rd	0010011	I	slti
imm[11:0]		rs1	011	rd	0010011	I	sltiu
imm[11:0]		rs1	100	rd	0010011	I	xori
imm[11:0]		rs1	110	rd	0010011	I	ori
imm[11:0]		rs1	111	rd	0010011	I	andi
0000000	shamt	rs1	001	rd	0010011	I	slli
0000000	shamt	rs1	101	rd	0010011	I	srli
0100000	shamt	rs1	101	rd	0010011	I	srai
0000000	rs2	rs1	000	rd	0110011	R	add
0100000	rs2	rs1	000	rd	0110011	R	sub
0000000	rs2	rs1	001	rd	0110011	R	sll
0000000	rs2	rs1	010	rd	0110011	R	slt
0000000	rs2	rs1	011	rd	0110011	R	sltu
0000000	rs2	rs1	100	rd	0110011	R	xor
0000000	rs2	rs1	101	rd	0110011	R	srl
0100000	rs2	rs1	101	rd	0110011	R	sra
0000000	rs2	rs1	110	rd	0110011	R	or
0000000	rs2	rs1	111	rd	0110011	R	and
0000	pred	succ	00000	000	00000	0001111	I	fence
0000	0000	0000	00000	001	00000	0001111	I	fence.i
000000000000		00000	000	00000	1110011	I	ecall
000000000001		00000	000	00000	1110011	I	ebreak
csr		rs1	001	rd	1110011	I	csrrw
csr		rs1	010	rd	1110011	I	csrrs
csr		rs1	011	rd	1110011	I	csrrc
csr		zimm	101	rd	1110011	I	csrrwi
csr		zimm	110	rd	1110011	I	csrrsi
csr		zimm	111	rd	1110011	I	csrrci

Figure 2.3: RV32I opcode map has instruction layout, opcodes, format type, and names. (Table 19.2 of [Waterman and Asanović 2017] is the basis of this figure.)

only a two-operand instruction, the compiler or assembly language programmer must use an extra move instruction to preserve the destination operand. Third, in RISC-V the specifiers of the registers to be read and written are always in the same location in all instructions, which means the register accesses can begin before decoding the instruction. Many other ISAs reuse a field as a source in some instructions and as a destination in others (e.g., ARM-32 and MIPS-32), which forces addition of extra hardware to be placed in a potentially time-critical path to select the proper field. Fourth, the immediate fields in these formats are always sign extended, and the sign bit is always in the most significant bit of the instruction. This decision means sign extension of the immediate, which may also be on a critical timing path, can proceed before decoding the instruction.

■ *Elaboration: B- and J-type formats?*

As mentioned below, the immediate field is rotated 1 bit for branch instructions, a variation of the S format that we relabel the B format. The immediate field of jump instructions rotated 12 bits for jump instructions, a variation of the U format relabeled J format. Hence, there are really four basic formats, but we can conservatively count RISC-V as having six formats.

To help programmers, a bit pattern of all zeros is an illegal RV32I instruction. Thus, erroneous jumps into zeroed memory regions will immediately trap, an aid to debugging. Similarly, the bit pattern of all ones is an illegal instruction, which will trap other common errors such as unprogrammed non-volatile memory devices, disconnected memory buses, or broken memory chips.

To leave ample room for ISA extensions, the base RV32I ISA uses less than ⅛-th of the encoding space in the 32-bit instruction word. The architects also carefully picked the RV32I opcodes so that instructions with common datapath operations share as many of the same opcode bit values as possible, which simplifies the control logic. Finally, as we shall see, the branch and jump addresses in the B and J formats must be shifted left 1 bit so as to multiply the addresses by 2, thereby giving branches and jumps a greater range. RISC-V rotates the bits in the immediate operands from a natural placement to reduce the instruction signal fanout and immediate multiplexing cost by almost a factor for two, which again simplifies datapath logic on low-end implementations.

What's Different? We'll end each section in this and following chapters with description on how RISC-V differs from other ISAs. The contrast is often what RISC-V is missing. Architects demonstrate good taste by the features they omit as well as by what they include.

The ARM-32 12-bit immediate field is not simply a constant but an input to a function that produces a constant: 8 bits are zero-extended to full width and then rotated right by the value in the 4 remaining bits multiplied by 2. The hope was encoding more useful constants in 12 bits would reduce the number of executed instructions. ARM-32 also dedicates a precious four bits in most instruction formats to conditional execution. Despite being infrequently used, conditional execution adds to the complexity of *out-of-order processors*.

Sign-extended immediates even help logical instructions. For example, x & 0xfffffff0 uses only the single instruction andi in RISC-V, but it requires two instructions in MIPS-32 (addiu to load the constant, then and), since MIPS zero-extends logical immediates. ARM-32 needed to add an additional instruction, bic, that performs rx & immediate to compensate for zeroextending immediates.

Programmability

Room for Growth

Cost

RISC-V implementations all use the same opcodes for the optional extensions such as RV32M, RV32F, and so on. Non-standard extensions that are unique to processor are restricted to a reserved opcode space in RISC-V.

Performance

■ *Elaboration: Out-of-order processors*

are high-speed, pipelined processors that execute instructions opportunistically instead of in lock-step program order. A critical feature of such processors is *register renaming*, which maps the register names in the program onto a larger number of internal physical registers. The problem with conditional execution is that the new physical register must be written whether or not the condition holds, so the old value of the destination register must be read as a third operand, to be copied to the new destination register in case the condition didn't hold. The extra operand increases the cost of the register file, register renamer, and out-of-order execution hardware.

2.3   RV32I Registers

Programmability

Simplicity

Pipelining is used by all but the cheapest processors today to get good performance. Like an industrial assembly line, they get higher throughput by overlapping the execution of many instructions at once. To pull this off, the processors predict branch outcomes, which they can do with more than 90% accuracy. When they mispredict, they re-execute instructions. Early microprocessors had a 5-stage pipeline, which meant 5 instructions overlapped execution. Recent ones have more than 10 pipeline stages. The ARM v8 successor to the ARM-32 dropped the PC as a general-purpose register, effectively admitting that it was a mistake.

Simplicity

Figure 2.4 lists the RV32I registers and their names as determined by the RISC-V application binary interface (ABI). We will use the ABI names in our code examples to make them easier to read. To the joy of assembly language programmers and compiler writers, RV32I has 31 registers plus x0, which always has the value 0. ARM-32 has merely 16 registers while x86-32 has only 8!

What's Different? Dedicating a register to zero is a surprisingly large factor in simplifying the RISC-V ISA. Figure 3.3 on page 36 in Chapter 3 gives many examples of operations that are native instructions in ARM-32 and x86-32, which don't have a zero register, but can be synthesized from RV32I instructions simply by using the zero register as an operand.

The PC is one of ARM-32's 16 registers, which means that any instruction that changes a register may also as a side effect be a branch instruction. The PC as a register complicates hardware branch prediction, whose accuracy is vital for good pipelined-performance, since every instruction might be a branch instead of 10–20% of instructions executed in programs for typical ISAs. It also means one less general-purpose register.

2.4   RV32I Integer Computation

Appendix A gives details of all of the RISC-V instructions, including formats and opcodes. In this section, and similar sections of the following chapters, we give an ISA overview that should be sufficient for knowledgeable assembly language programmers, as well as highlight the features that demonstrate the seven ISA metrics from Chapter 1.

The simple arithmetic instructions (add, sub), logical instructions (and, or, xor), and shift instructions (sll, srl, sra) in Figure 2.1 are just as you would expect in any ISA. They read two 32-bit values from registers and write a 32-bit result to the destination register. RV32I also offers immediate versions of these instructions. Unlike ARM-32, immediates are always sign-extended so that they can be negative when needed, which is why there is no need for an immediate version of sub.

Programs can generate a Boolean value from the result of a comparison. To accommodate such cases, RV32I offers a *set less than* instruction, which sets the destination register to 1 if the first operand is less than the second, or 0 otherwise. As one would expect, there is a signed version (slt) and an unsigned version (sltu) for signed and unsigned integers as well as immediate versions for both (slti, sltiu). As we shall see, while RV32I branches can check for all relationships between two registers, some conditional expressions involve relationships between many pairs of registers. The compiler or assembly language programmer

31	0	
x0 / zero		Hardwired zero
x1 / ra		Return address
x2 / sp		Stack pointer
x3 / gp		Global pointer
x4 / tp		Thread pointer
x5 / t0		Temporary
x6 / t1		Temporary
x7 / t2		Temporary
x8 / s0 / fp		Saved register, frame pointer
x9 / s1		Saved register
x10 / a0		Function argument, return value
x11 / a1		Function argument, return value
x12 / a2		Function argument
x13 / a3		Function argument
x14 / a4		Function argument
x15 / a5		Function argument
x16 / a6		Function argument
x17 / a7		Function argument
x18 / s2		Saved register
x19 / s3		Saved register
x20 / s4		Saved register
x21 / s5		Saved register
x22 / s6		Saved register
x23 / s7		Saved register
x24 / s8		Saved register
x25 / s9		Saved register
x26 / s10		Saved register
x27 / s11		Saved register
x28 / t3		Temporary
x29 / t4		Temporary
x30 / t5		Temporary
x31 / t6		Temporary

32

31	0
pc	

32

Figure 2.4: The registers of RV32I. Chapter 3 explains the RISC-V calling convention, the rationale behind the various pointers (sp, gp, tp, fp), Saved registers (s0-s11), and Temporaries (t0-t6). (Figure 2.1 and Table 20.1 of [Waterman and Asanović 2017] is the basis of this figure.)

could use slt and the logical instructions and, or, xor to resolve more elaborate conditional expressions.

Programmability

The two remaining integer computation instructions in Figure 2.1 help with assembly and linking. *Load upper immediate* (lui) loads a 20-bit constant into the most significant 20 bits of a register. It can be followed by a standard immediate instruction to create a 32-bit constant from only two 32-bit RV32I instructions. *Add upper immediate to PC* (auipc) supports two-instruction sequences to access arbitrary offsets from the PC for both control-flow transfers and data accesses. The combination of an auipc and the 12-bit immediate in a jalr (see below) can transfer control to any 32-bit PC-relative address, while an auipc plus the 12-bit immediate offset in regular load or store instructions can access any 32-bit PC-relative data address.

Simplicity

What's Different? First, there are no byte or half-word integer computation operations. The operations are always the full register width. Memory accesses take orders of magnitude more energy than arithmetic operations, so narrow data accesses can save significant energy, but narrow operations do not. ARM-32 has the unusual feature of having an option to shift one of the operands in most arithmetic-logic operations, which complicates the datapath and is rarely needed [Hohl and Hinds 2016]; RV32I has separate shift instructions.

The ARM v8 successor ISA to ARM-32 dropped the optional shift operation from ALU instructions, again suggesting that it was a mistake for ARM-32.

Nor does RV32I include multiply and divide; they comprise the optional RV32M extension (see Chapter 4). Unlike ARM-32 and x86-32, the full RISC-V software stack can run without them, which can shrink embedded chips. While not a hardware issue, the MIPS-32 *assembler* may replace a multiply with a sequence shifts and adds to try to improve performance, which may confuse the programmer seeing instructions executed not found in the assembly language program. RV32I also omits rotate instructions and detection of integer arithmetic overflow. Both can be calculated in a few RV32I instructions (see Section 2.6).

Cost

■ *Elaboration: "Bit twiddling" instructions*

such as rotate are under consideration by the RISC-V Foundation as part of an optional instruction extension called RV32B (see Chapter 11).

■ *Elaboration: xor enables a magic trick.*

You can exchange two values without using an intermediate register! This code exchanges the values of x1 and x2. We leave the proof to the reader. Hint: exclusive or is commutative ($a \oplus b = b \oplus a$), associative (($a \oplus b) \oplus c = a \oplus (b \oplus c)$), is its own inverse ($a \oplus a = 0$), and has an identity ($a \oplus 0 = a$).

```
xor   x1,x1,x2 # x1'  == x1^x2, x2'  == x2
xor   x2,x1,x2 # x1'  == x1^x2, x2'  == x1'^x2 == x1^x2^x2 == x1
xor   x1,x1,x2 # x1'' == x1'^x2' == x1^x2^x1 == x1^x1^x2 == x2, x2'  == x1
```

However fascinating, RISC-V's ample register set usually lets compilers find a scratch register, so it rarely uses the XOR-swap.

2.5   RV32I Loads and Stores

As well as providing loads and stores of 32-bit words (lw, sw), Figure 2.1 shows that RV32I has loads for signed and unsigned bytes and halfwords (lb, lbu, lh, lhu) and stores for bytes and halfwords (sb, sh). Signed bytes and halfwords are sign-extended to 32 bits and written to the destination registers. This widening of narrow data allows subsequent integer computation instructions to operate correctly on all 32 bits, even if the natural data types are narrower. Unsigned bytes and halfwords, useful for text and unsigned integers, are zero-extended to 32 bits before being written to the destination register.

The only addressing mode for loads and stores is adding a sign-extended 12-bit immediate to a register, called displacement addressing mode in x86-32 [Irvine 2014].

What's Different? RV32I omitted the sophisticated addressing modes of ARM-32 and x86-32. Alas, all ARM-32 addressing modes aren't available for all data types, but RV32I addressing does not discriminate against any data type. RISC-V can imitate some x86 addressing modes. For example, setting the immediate field to 0 has the same effect as the register-indirect addressing mode. Unlike x86-32, RISC-V has no special stack instructions. By using one of the 31 registers as the stack pointer (see Figure 2.4), the standard addressing mode gets most of the benefits of push and pop instructions without the added ISA complexity. Unlike MIPS-32, RISC-V rejected *delayed load*. Similar in spirit to delayed branches, MIPS-32 redefined the load so the data is unavailable until two instructions later, when it would show up in a five-stage pipeline. Whatever benefit it had evaporated for the longer pipelines that came later.

While ARM-32 and MIPS-32 require data to be aligned naturally to data-sized boundaries in memory, RISC-V does not. Misaligned accesses are sometimes required when porting legacy code. One option is to disallow misaligned accesses in the base ISA and then provide some separate instructions support for misaligned accesses, such as Load Word Left and Load Word Right of MIPS-32. This option would complicate register access, however, since lwl and lwr require writing pieces of registers instead of simply full registers. Requiring instead that the regular loads and stores support misaligned accesses simplified the overall design.

■ *Elaboration: Endianness*

RISC-V chose *little-endian* byte ordering because it is dominant commercially: all x86-32 systems, and Apple iOS, Google Android OS, and Microsoft Windows for ARM are all little-endian. Since the endian order matters only when accessing the identical data both as a word and as bytes, endianness affects few programmers.

Cost

2.6 RV32I Conditional Branch

RV32I can compare two registers and branch on the result if they are equal (beq), not equal (bne), greater than or equal (bge), or less than (blt). The latter two cases are signed comparisons, but RV32I also offers unsigned versions: bgeu and bltu. The two remaining relationships (greater than and less than or equal) can be checked simply by reversing the operands, since $x < y$ means that $y > x$ and $x \geq y$ implies $y \leq x$.

Since RISC-V instructions must be a multiple of two bytes long—see Chapter 7 to learn about the optional 2-byte instructions—the branch addressing mode multiplies the 12-bit immediate by 2, sign-extends it, and then adds it to the PC. PC-relative addressing helps with position independent code and thereby reduces the work of the linker and loader (Chapter 3).

What's Different? As noted above, RISC-V excluded the infamous delayed branch of MIPS-32, Oracle SPARC, and others. It also avoided the condition codes of ARM-32 and x86-32 for conditional branches. They add extra state that is implicitly set by most instructions, which needlessly complicate the dependence calculation for out-of-order execution.

bltu **allows signed array bounds to be checked with a single instruction**, since any negative index will compare greater than any nonnegative bound!

Programmability

Finally, it omitted the loop instructions of the x86-32: `loop`, `loope`, `loopz`, `loopne`, `loopnz`.

Simplicity

■ *Elaboration: Multiword addition without condition codes*

is done as follows in RV32I by using `sltu` to calculate the carry-out:

```
add  a0,a2,a4 # add lower 32 bits: a0 = a2 + a4
sltu a2,a0,a2 # a2' = 1 if (a2+a4) < a2, a2' = 0 otherwise
add  a5,a3,a5 # add upper 32 bits: a5 = a3 + a5
add  a1,a2,a5 # add carry-out from lower 32 bits
```

■ *Elaboration: Reading the PC*

The current PC can be obtained by setting the U-immediate field of `auipc` to 0. For the x86-32, to read the PC you need to call a function (which pushes the PC to the stack); the callee then reads the pushed PC from the stack, and finally returns the PC (by popping the stack). So reading the current PC took 1 store, 2 loads, and 2 taken jumps!

■ *Elaboration: Software checking of overflow*

Most but not all programs ignore integer arithmetic overflow, so RISC-V relies on software overflow checking. Unsigned addition requires only a single additional branch instruction after the addition: `addu t0, t1, t2; bltu t0, t1, overflow`.
For signed addition, if one operand's sign is known, overflow checking requires only a single branch after the addition: `addi t0, t1, +imm; blt t0, t1, overflow`.
This covers the common case of addition with an immediate operand. For general signed addition, three additional instructions after the addition are required, observing that the sum should be less than one of the operands if and only if the other operand is negative.

```
add t0, t1, t2
slti t3, t2, 0       # t3 = (t2<0)
slt t4, t0, t1       # t4 = (t1+t2<t1)
bne t3, t4, overflow # overflow if (t2<0) && (t1+t2>=t1)
                     #             || (t2>=0) && (t1+t2<t1)
```

2.7   RV32I Unconditional Jump

Simplicity

The single *jump and link* instruction (`jal`) in Figure 2.1 serves dual functions. To support procedure calls, it saves the address of the next instruction PC+4 into the destination register, normally the return address register `ra` (see Figure 2.4). To support unconditional jumps, we use the zero register (x0) instead of `ra` as the destination register, as x0 can't be changed. Like branches, `jal` multiplies its 20-bit branch address by 2, sign extends it, and then adds the result to the PC to form the jump address.

The register version of jump and link (`jalr`) is similarly multipurpose. It can call a procedure to a dynamically calculated address or simply perform a procedure return by selecting the `ra` as the source register, and the zero register (x0) again as the destination register. Switch or case statements, which calculate a jump address, can also use `jalr` with the zero register as the destination register.

```
void insertion_sort(long a[], size_t n)
{
  for (size_t i = 1, j; i < n; i++) {
    long x = a[i];
    for (j = i; j > 0 && a[j-1] > x; j--) {
      a[j] = a[j-1];
    }
    a[j] = x;
  }
}
```

Figure 2.5: Insertion Sort in C. While simple, Insertion Sort has many advantages over complicated sorting algorithms: it is memory efficient and fast for small data sets while being adaptive, stable, and online. GCC compilers produced the code for the following four figures. We set the optimization flags to reduce code size, as that produced the easiest to understand code.

What's Different? RV32I shunned intricate procedure call instructions, such as the `enter` and `leave` instructions of the x86-32, or *register windows* as found in the Intel Itanium, Oracle SPARC, and Cadence Tensilica.

> **Register windows** accelerated function call by having many more registers than 32. A new function would get a new set or *window* of 32 registers on a call. To pass arguments, the windows overlapped, meaning some registers were in two adjacent windows.

2.8 RV32I Miscellaneous

The Control Status Register instructions (`csrrc`, `csrrs`, `csrrw`, `csrrci`, `csrrsi`, `csrrwi`) in Figure 2.1 provide easy access to registers that help measure program performance. These 64-bit counters, which can be read 32 bits at a time, measure wall clock time, clock cycles executed, and number of instructions retired.

The `ecall` instruction makes requests to the supporting execution environment, such as system calls. Debuggers use the `ebreak` instruction to transfer control to a debugging environment.

The `fence` instruction sequences device I/O and memory accesses as viewed by other threads and by external devices or coprocessors. The `fence.i` instruction synchronizes the instruction and data streams. RISC-V does not guarantee that stores to instruction memory are visible to instruction fetches in the same processor until a `fence.i` instruction executes.

Chapter 10 covers the RISC-V system instructions.

What's Different? RISC-V uses memory mapped I/O instead of the in, ins, insb, insw and out, outs, outsb, outsw instructions of the x86-32. It supports strings using byte loads and stores instead of the 16 special string instructions of the x86-32 rep, movs, coms, scas, lods, ...

Simplicity

2.9 Comparing RV32I, ARM-32, MIPS-32, and x86-32 using Insertion Sort

We've introduced the RISC-V base instruction set, and commented upon its choices as compared to ARM-32, MIPS-32, and x86-32. We'll now do a head-to-head comparison. Figure 2.5 shows Insertion Sort in C, which will be our benchmark. Figure 2.6 is a table that summarizes the number of instructions and number of bytes in Insertion Sort for the ISAs.

ISA	ARM-32	ARM Thumb-2	MIPS-32	microMIPS	x86-32	RV32I	RV32I+RVC
Instructions	19	18	24	24	20	19	19
Bytes	76	46	96	56	45	76	52

Figure 2.6: Number of instructions and code size for Insertion Sort for these ISAs. Chapter 7 describes ARM Thumb-2, microMIPS, and RV32C.

We move the code examples to after the end of the chapter text to maintain the flow of the writing in this and following chapters.

Figures 2.8 to 2.11 show the compiled code for RV32I, ARM-32, MIPS-32, and x86-32. Despite the emphasis on simplicity, the RISC-V version uses the same or fewer instructions, and the code sizes of the architectures are quite close. In this example, the compare-and-execute branches of RISC-V save as many instructions as do the fancier address modes and the push and pop instructions of ARM-32 and x86-32 in Figures 2.9 and 2.11.

2.10   Concluding Remarks

Those who cannot remember the past are condemned to repeat it.

—George Santayana, 1905

The genealogy of all RISC-V instructions is chronicled in [Chen and Patterson 2016].

Figure 2.7 uses the seven metrics of ISA design from Chapter 1 to organize the lessons from past ISAs listed it the previous sections, and shows the positive outcomes for RV32I. We're *not* implying that RISC-V is the first ISA to have those outcomes. Indeed, RV32I inherits the following from RISC-I, its great-great-grandparent [Patterson 2017]:

The Lindy effect [Lin 2017] observes that the future life expectancy of a technology or idea is proportional to its age. It has stood the test of time, so the longer it has survived in the past, the longer it likely will survive in the future. If that hypothesis holds, RISC architecture may be a good idea for a long time.

- 32-bit byte-addressable address space

- All instructions are 32-bit long

- 31 registers, all 32 bits wide, with register 0 hardwired to zero

- All operations are between registers (none are register-to-memory)

- Load/store word plus signed and unsigned load/store byte and halfword

- Immediate option for all arithmetic, logical, and shift instructions

- Immediates always sign-extend

- One data addressing mode (register + immediate) and PC-relative branching

- No multiply or divide instructions

- An instruction to load a wide immediate into the upper part of register so that a 32-bit constant takes only two instructions

Elegance

RISC-V benefits from starting one-quarter to one-third century later, which allowed its architects to follow Santayana's advice to borrow the good ideas but to not repeat the mistakes of the past—including those of RISC-I—in the current RISC-V ISA. Moreover, the RISC-V Foundation will grow the ISA slowly via optional extensions to prevent the rampant incrementalism that has plagued successful ISAs of the past.

| | Mistakes of the Past | | | Lessons learned |
	ARM-32 (1986)	MIPS-32 (1986)	x86-32 (1978)	RV32I (2011)
Cost	Integer multiply mandatory	Integer multiply and divide mandatory	8-bit and 16-bit operations. Integer multiply and divide mandatory	No 8-bit and 16-bit operations. Integer multiply and divide optional (RV32M)
Simplicity	No zero register. Conditional instruction execution. Complex data address modes. Stack instructions (push/pop). Shift-option for arithmetic/logic instructions	Zero- and sign-extended immediates. Some arithmetic instructions can cause overflow traps	No zero register. Complex procedure call/return instructions (enter/leave). Stack instructions (push/pop). Complex data address modes. Loop instructions	Register x0 dedicated to 0. Immediates only sign-extended. One data addressing mode. No conditional execution. No complex call/return or stack instructions. No traps for arithmetic overflow. Separate shift instructions
Performance	Condition codes for branches. Source and destination registers vary in instruction format. Load multiple. Computed immediates. PC a general purpose register	Source and destination registers vary in instruction format.	Condition codes for branches. At most 2 registers per instruction	Compare and branch instructions (no condition codes). 3 registers per instruction. No load multiple. Source and destination registers fixed in instruction format. Constant immediates. PC not a general purpose register
Isolate architecture from implementation	Exposes the pipeline length when writing the PC as a general purpose register	Delayed branch. Delayed load. HI and LO registers just for multiply and divide	Registers not general purpose (AX, CX, DX, DI, SI have unique uses)	No delayed branch. No delayed load. General purpose registers
Room for growth	Limited available opcode space	Limited available opcode space		Generous available opcode space
Program size	Only 32-bit instructions (+Thumb-2 as separate ISA)	Only 32-bit instructions (+microMIPS as separate ISA)	Byte-variable instructions, but poor choices	32-bit instructions + 16-bit RV32C extension
Ease of programming / compiling / linking	Only 15 registers. Aligned data in memory. Irregular data address modes. Inconsistent performance counters	Aligned data in memory. Inconsistent performance counters	Only 8 registers. No PC-relative data addressing. Inconsistent performance counters	31 registers. Data can be unaligned. PC-relative data addressing. Symmetric data address mode. Performance counters defined in architecture

Figure 2.7: Lessons that RISC-V architects learned from past instruction set mistakes. Often the lesson was simply to avoid ISA "optimizations" of the past. The lessons and mistakes are classified by the seven ISA metrics from Chapter 1. Many features listed under cost, simplicity, and performance could be swapped with each other, as it's a matter of taste, but they are important no matter where they appear.

■ *Elaboration: Is RV32I unique?*

Early microprocessors had separate chips for floating-point arithmetic, so those instructions were optional. Moore's Law soon brought everything on chip, and modularity faded in ISAs. Subsetting the full ISA in simpler processors and trapping to software to emulate them goes back decades, with the IBM 360 model 44 and the Digital Equipment microVAX as examples. RV32I is different in that the full software stack needs only the base instructions, so an RV32I processor need not trap repeatedly for omitted instructions in RV32G. Probably the closest ISA to RISC-V in that respect is the Tensilica Xtensa, which is aimed at embedded applications. Its 80-instruction base ISA is intended to be extended by users with custom instructions that accelerate their applications. RV32I has a simpler base ISA, has a 64-bit address version, and offers extensions that target supercomputers as well as microcontrollers.

2.11   To Learn More

Lindy effect, 2017. URL https://en.wikipedia.org/wiki/Lindy_effect.

T. Chen and D. A. Patterson. RISC-V genealogy. Technical Report UCB/EECS-2016-6, EECS Department, University of California, Berkeley, Jan 2016. URL http://www2.eecs.berkeley.edu/Pubs/TechRpts/2016/EECS-2016-6.html.

W. Hohl and C. Hinds. *ARM Assembly Language: Fundamentals and Techniques.* CRC Press, 2016.

K. R. Irvine. *Assembly language for x86 processors.* Prentice Hall, 2014.

D. Patterson. How close is RISC-V to RISC-I?, 2017.

A. Waterman and K. Asanović, editors. *The RISC-V Instruction Set Manual, Volume I: User-Level ISA, Version 2.2.* May 2017. URL https://riscv.org/specifications/.

Notes

[1]http://parlab.eecs.berkeley.edu

```
# RV32I (19 instructions, 76 bytes, or 52 bytes with RVC)
# a1 is n, a3 points to a[0], a4 is i, a5 is j, a6 is x
    0: 00450693   addi a3,a0,4      # a3 is pointer to a[i]
    4: 00100713   addi a4,x0,1      # i = 1
Outer Loop:
    8: 00b76463   bltu a4,a1,10     # if i < n, jump to Continue Outer loop
Exit Outer Loop:
    c: 00008067   jalr x0,x1,0      # return from function
Continue Outer Loop:
   10: 0006a803   lw   a6,0(a3)     # x = a[i]
   14: 00068613   addi a2,a3,0      # a2 is pointer to a[j]
   18: 00070793   addi a5,a4,0      # j = i
Inner Loop:
   1c: ffc62883   lw   a7,-4(a2)    # a7 = a[j-1]
   20: 01185a63   bge  a6,a7,34     # if a[j-1] <= a[i], jump to Exit Inner Loop
   24: 01162023   sw   a7,0(a2)     # a[j] = a[j-1]
   28: fff78793   addi a5,a5,-1     # j--
   2c: ffc60613   addi a2,a2,-4     # decrement a2 to point to a[j]
   30: fe0796e3   bne  a5,x0,1c     # if j != 0, jump to Inner Loop
Exit Inner Loop:
   34: 00279793   slli a5,a5,0x2    # multiply a5 by 4
   38: 00f507b3   add  a5,a0,a5     # a5 is now byte address oi a[j]
   3c: 0107a023   sw   a6,0(a5)     # a[j] = x
   40: 00170713   addi a4,a4,1      # i++
   44: 00468693   addi a3,a3,4      # increment a3 to point to a[i]
   48: fc1ff06f   jal  x0,8         # jump to Outer Loop
```

Figure 2.8: RV32I code for Insertion Sort in Figure 2.5. The address in hexadecimal is on the left, the machine language code in hexadecimal is next, and then the assembly language instruction followed by a comment. RV32I allocates two registers to point to a[j] and a[j-1]. It has plenty of registers, some of which the ABI sets aside for procedure calls. Unlike the other ISAs, it skips saving and restoring these registers to memory. While the code size is larger than x86-32, using the optional RV32C instructions (see Chapter 7) closes the size gap. Note the compare and branch instructions avoid the three compare instructions that ARM-32 and x86-32 require.

```
# ARM-32 (19 instructions, 76 bytes; or 18 insns/46 bytes with Thumb-2)
# r0 points to a[0], r1 is n, r2 is j, r3 is i, r4 is x
    0: e3a03001 mov  r3, #1                # i = 1
    4: e1530001 cmp  r3, r1                # i vs. n (unnecessary?)
    8: e1a0c000 mov  ip, r0                # ip = a[0]
    c: 212fff1e bxcs lr                    # don't let return address change ISAs
   10: e92d4030 push {r4, r5, lr}          # save r4, r5, return address
Outer Loop:
   14: e5bc4004 ldr  r4, [ip, #4]!         # x = a[i] ; increment ip
   18: e1a02003 mov  r2, r3                # j = i
   1c: e1a0e00c mov  lr, ip                # lr = a[0] (using lr as scratch reg)
Inner Loop:
   20: e51e5004 ldr  r5, [lr, #-4]         # r5 = a[j-1]
   24: e1550004 cmp  r5, r4                # compare a[j-1] vs. x
   28: da000002 ble  38                    # if a[j-1]<=a[i], jump to Exit Inner Loop
   2c: e2522001 subs r2, r2, #1            # j--
   30: e40e5004 str  r5, [lr], #-4         # a[j] = a[j-1]
   34: 1afffff9 bne  20                    # if j != 0, jump to Inner Loop
Exit Inner Loop:
   38: e2833001 add  r3, r3, #1            # i++
   3c: e1530001 cmp  r3, r1                # i vs. n
   40: e7804102 str  r4, [r0, r2, lsl #2]  # a[j] = x
   44: 3afffff2 bcc  14                    # if i < n, jump to Outer Loop
   48: e8bd8030 pop  {r4, r5, pc}          # restore r4, r5, and return address
```

Figure 2.9: ARM-32 code for Insertion Sort in Figure 2.5. The address in hexadecimal is on the left, the machine language code in hexadecimal is next, and then the assembly language instruction followed by a comment. Short on registers, ARM-32 saves two of them on the stack for later reuse along with the return address. It uses an addressing mode that scales i and j to be byte addresses. Given that a branch has the potential to change ISAs between ARM-32 and Thumb-2, bxcs first sets the least significant bit of the return address to 0 before saving it. The condition codes save one compare instruction to check j after decrementing it, but there are still three compares elsewhere.

```
# MIPS-32 (24 instructions, 96 bytes, or 56 bytes with microMIPS)
# a1 is n, a3 is pointer to a[0], v0 is j, v1 is i, t0 is x
    0: 24860004 addiu a2,a0,4    # a2 is pointer to a[i]
    4: 24030001 li    v1,1       # i = 1
Outer Loop:
    8: 0065102b sltu  v0,v1,a1   # set on i < n
    c: 14400003 bnez  v0,1c      # if i<n, jump to Continue Outer Loop
   10: 00c03825 move  a3,a2      # a3 is pointer to a[j] (slot filled)
   14: 03e00008 jr    ra         # return from function
   18: 00000000 nop              # branch delay slot unfilled
Continue Outer Loop:
   1c: 8cc80000 lw    t0,0(a2)   # x = a[i]
   20: 00601025 move  v0,v1      # j = i
Inner Loop:
   24: 8ce9fffc lw    t1,-4(a3)  # t1 = a[j-1]
   28: 00000000 nop              # load delay slot unfilled
   2c: 0109502a slt   t2,t0,t1   # set a[i] < a[j-1]
   30: 11400005 beqz  t2,48      # if a[j-1]<=a[i], jump to Exit Inner Loop
   34: 00000000 nop              # branch delay slot unfilled
   38: 2442ffff addiu v0,v0,-1   # j--
   3c: ace90000 sw    t1,0(a3)   # a[j] = a[j-1]
   40: 1440fff8 bnez  v0,24      # if j != 0, jump to Inner Loop
   44: 24e7fffc addiu a3,a3,-4   # decr. a2 to point to a[j] (slot filled)
Exit Inner Loop:
   48: 00021080 sll   v0,v0,0x2  #
   4c: 00821021 addu  v0,a0,v0   # v0 now byte address oi a[j]
   50: ac480000 sw    t0,0(v0)   # a[j] = x
   54: 24630001 addiu v1,v1,1    # i++
   58: 1000ffeb b     8          # jump to Outer Loop
   5c: 24c60004 addiu a2,a2,4    # incr. a2 to point to a[i] (slot filled)
```

Figure 2.10: MIPS-32 code for Insertion Sort in Figure 2.5. The address in hexadecimal is on the left, the machine language code in hexadecimal is next, and then the assembly language instruction followed by a comment. The MIPS-32 code has three nop instructions, which adds to its length. Two are due to delayed branches and one is due to the delayed load. The compiler couldn't find useful instructions to place in those delay slots. The delayed branches also make the code harder to understand, since the instruction that follows is also executed when a branch or jump is taken. For example, the last instruction (addiu) at address 5c is part of the loop even though it trails the branch instruction.

```
# x86-32 (20 instructions, 45 bytes)
# eax is j, ecx is x, edx is i
# pointer to a[0] is in memory at address esp+0xc, n is in memory at esp+0x10
   0: 56              push esi              # save esi on stack (esi needed below)
   1: 53              push ebx              # save ebx on stack (ebx needed below)
   2: ba 01 00 00 00  mov  edx,0x1          # i = 1
   7: 8b 4c 24 0c     mov  ecx,[esp+0xc]    # ecx is pointer to a[0]
Outer Loop:
   b: 3b 54 24 10     cmp  edx,[esp+0x10]   # compare i vs. n
   f: 73 19           jae  2a <Exit Loop>   # if i >= n, jump to Exit Outer Loop
  11: 8b 1c 91        mov  ebx,[ecx+edx*4]  # x = a[i]
  14: 89 d0           mov  eax,edx          # j = i
Inner Loop:
  16: 8b 74 81 fc     mov  esi,[ecx+eax*4-0x4] # esi = a[j-1]
  1a: 39 de           cmp  esi,ebx          # compare a[j-1] vs. x
  1c: 7e 06           jle  24 <Exit Loop>   # if a[j-1]<=a[i],jump Exit Inner Loop
  1e: 89 34 81        mov  [ecx+eax*4],esi  # a[j] = a[j-1]
  21: 48              dec  eax              # j--
  22: 75 f2           jne  16 <Inner Loop>  # if j != 0, jump to Inner Loop
Exit Inner Loop:
  24: 89 1c 81        mov  [ecx+eax*4],ebx  # a[j] = x
  27: 42              inc  edx              # i++
  28: eb e1           jmp  b <Outer Loop>   # jump to Outer Loop
Exit Outer Loop:
  2a: 5b              pop  ebx              # restore old value of ebx from stack
  2b: 5e              pop  esi              # restore old value of esi from stack
  2c: c3              ret                   # return from function
```

Figure 2.11: x86-32 code for Insertion Sort in Figure 2.5. The address in hexadecimal is on the left, the machine language code in hexadecimal is next, and then the assembly language instruction followed by a comment. Lacking registers, the x86-32 saves two of them on the stack. Moreover, two of the variables allocated to registers in RV32I are instead kept in memory (n and the pointer to a[0]). It uses the Scaled Indexed addressing mode to good effect for accessing a[i] and a[j]. Seven of the 20 x32-86 instructions are one byte long, which gives the x86-32 good code size for this simple program. There are two popular versions of x86 assembly language: Intel/Microsoft and AT&T/Linux. We use the Intel syntax, in part because it matches the operand order of RISC-V, ARM-32, and MIPS-32 with the destination on the left and the source(s) on the right, while the operands are vice versa for AT&T (and the registers prepend a % before their names). This seemingly trivial matter is nearly a religious issue for some programmers. Pedagogy drives our choice, not orthodoxy.

3 RISC-V Assembly Language

Ivan Sutherland (1938-) is called the father of computer graphics for the invention of Sketchpad—the 1962 forerunner of the graphical user interface in modern computing—which led to a Turing Award.

It's very satisfying to take a problem we thought difficult and find a simple solution. The best solutions are always simple.

—Ivan Sutherland

3.1 Introduction

Figure 3.1 shows the four classic steps in translation starting from a C program and ending with a machine-language program ready to run in the computer. This chapter covers the last three steps, but we begin with the role the assembler plays in the RISC-V calling convention.

3.2 Calling convention

There are six general stages in calling a function [Patterson and Hennessy 2017]:

1. Place the arguments where the function can access them.

2. Jump to the function (using RV32I's `jal`).

3. Acquire local storage resources the function needs, saving registers as required.

4. Perform the desired task of the function.

5. Place the function result value where the calling program can access it, restore any registers, and release any local storage resources.

6. Since a function can be called from several points in a program, return control to the point of origin (using `ret`).

To obtain good performance, try to keep variables in registers rather than memory, but on the other hand, avoid going to memory frequently to save and restore these registers.

RISC-V fortunately has enough registers to offer the best of both worlds: keep operands in registers yet reduce the need to save and restore them. The insight is to have some registers that are *not* guaranteed to be preserved across a function call, called *temporary registers*, and some that are, called *saved registers*. Functions that avoid calling other functions are called *leaf* functions. When a leaf function has only a few arguments and local variables, we can keep everything in registers without "spilling" any to memory. If these conditions hold, then

Performance

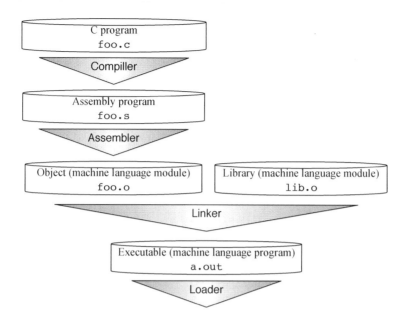

Figure 3.1: Steps of translation from C source code to a running program. These are the logical steps, although some steps are combined to accelerate translation. We use the Unix file suffix name convention for each type of file. The equivalent suffixes in MS-DOS are .C, .ASM, .OBJ, .LIB, **and** .EXE.

the program doesn't need to save register values in memory, and a surprising fraction of function calls fall into this happy case.

Other registers within a function call must be considered either in the same class as saved registers, which are preserved across a function call, or in the same class as the temporary registers, which are not. A function will change the register(s) containing the return value(s), so they are like temporary registers. There is no need to preserve the return address or the function arguments, so these registers also are like temporaries. The caller can rely on the remaining register, the stack pointer, to be unchanged by across a function call. Figure 3.2 lists the RISC-V application binary interface (ABI) names of registers and the convention on whether they are preserved or not across function calls.

Given the ABI conventions, we can see the standard RV32I code for function entry and exit. The function *prologue* looks like this:

```
entry_label:
  addi sp,sp,-framesize    # Allocate space for stack frame
                           # by adjusting stack pointer (sp register)
  sw   ra,framesize-4(sp)  # Save return address (ra register)
  # save other registers to stack if needed
  ... # body of the function
```

If there are too many function arguments and variables to fit in the registers, the prologue allocates space on the stack for the function *frame*, as it is called. After the task of the function is complete, the *epilogue* undoes the stack frame and returns to the point of origin:

Register	ABI Name	Description	Preserved across call?
x0	zero	Hard-wired zero	—
x1	ra	Return address	No
x2	sp	Stack pointer	Yes
x3	gp	Global pointer	—
x4	tp	Thread pointer	—
x5	t0	Temporary/alternate link register	No
x6–7	t1–2	Temporaries	No
x8	s0/fp	Saved register/frame pointer	Yes
x9	s1	Saved register	Yes
x10–11	a0–1	Function arguments/return values	No
x12–17	a2–7	Function arguments	No
x18–27	s2–11	Saved registers	Yes
x28–31	t3–6	Temporaries	No
f0–7	ft0–7	FP temporaries	No
f8–9	fs0–1	FP saved registers	Yes
f10–11	fa0–1	FP arguments/return values	No
f12–17	fa2–7	FP arguments	No
f18–27	fs2–11	FP saved registers	Yes
f28–31	ft8–11	FP temporaries	No

Figure 3.2: Assembler mnemonics for RISC-V integer and floating-point registers. RISC-V has enough registers that the ABI can allocate registers that procedures or methods are free to use without saving or restoring when they don't call other procedures or methods themselves. The registers preserved across a procedure call are also named *caller saved* versus *callee saved* for those that aren't. Chapter 5 explains the floating-point f registers. (Table 20.1 of [Waterman and Asanović 2017] is the basis of this figure.)

```
# restore registers from stack if needed
lw    ra,framesize-4(sp)   # Restore return address register
addi  sp,sp, framesize     # De-allocate space for stack frame
ret                        # Return to calling point
```

We'll see an example that follows this ABI shortly, but first we need to explain the remaining assembly tasks beyond turning the ABI register names into register numbers.

■ *Elaboration: The saved and temporary registers aren't contiguous*

to support RV32E, an embedded version of RISC-V that has only 16 registers (see Chapter 11). It simply uses register numbers x0 to x15, so some saved and temporary registers are in this range, and the rest are in the last 16 registers. RV32E is smaller, but has no compiler support yet, since it doesn't match RV32I.

3.3 Assembly

The input to this step in Unix is a file with the suffix .s, such as foo.s; for MS-DOS it is .ASM.

Simplicity

The job of the assembler step of Figure 3.1 is not simply to produce object code from the instructions that the processor understands, but to extend them to include operations useful for the assembly language programmer or the compiler writer. This category, based on clever configurations of regular instructions, is called *pseudoinstructions*. Figures 3.3 and 3.4 list the RISC-V pseudoinstructions, with those in the first figure all relying on register x0 to always be zero while those in the second list do not. For example, ret mentioned above is actually a pseudoinstruction that the assembler replaces with jalr x0, x1, 0 (see Figure 3.3). The majority of RISC-V pseudoinstructions depend on x0. As you can see, setting aside one of the 32 registers to be hardwired to zero greatly simplifies the RISC-V instruction set by providing many popular operations—such as jump, return, and branch on equal to zero—as pseudoinstructions.

Figure 3.5 shows the classic "Hello world" program in C. The compiler produces the assembly language output in Figure 3.6 using the calling convention in Figure 3.2 and the pseudoinstructions from Figures 3.3 and 3.4.

The "Hello world" program is typically the first program run on a newly designed processor. Architects traditionally consider running the operating system well enough to print "Hello world" as a strong sign that their new chip largely works. They email this output to their management and colleagues, and then they celebrate.

The commands that start with a period are *assembler directives*. They are commands to the assembler rather than code to be translated by it. They tell the assembler where to place code and data, specify text and data constants for use in the program, and so forth. Figure 3.9 shows the assembler directives of RISC-V. For Figure 3.6, the directives are:

- .text—Enter text section.

- .align 2—Align following code to 2^2 bytes.

- .globl main—Declare global symbol "main".

- .section .rodata—Enter read-only data section.

- .balign 4—Align data section to 4 bytes.

- .string ''Hello, %s!\n''—Create this null-terminated string.

- .string ''world''—Create this null-terminated string.

The assembler produces the object file in Figure 3.7 using the Executable and Linkable Format (ELF) standard format [TIS Committee 1995].

Pseudoinstruction	Base Instruction(s)	Meaning
nop	addi x0, x0, 0	No operation
neg rd, rs	sub rd, x0, rs	Two's complement
negw rd, rs	subw rd, x0, rs	Two's complement word
snez rd, rs	sltu rd, x0, rs	Set if $\neq$ zero
sltz rd, rs	slt rd, rs, x0	Set if $<$ zero
sgtz rd, rs	slt rd, x0, rs	Set if $>$ zero
beqz rs, offset	beq rs, x0, offset	Branch if $=$ zero
bnez rs, offset	bne rs, x0, offset	Branch if $\neq$ zero
blez rs, offset	bge x0, rs, offset	Branch if $\leq$ zero
bgez rs, offset	bge rs, x0, offset	Branch if $\geq$ zero
bltz rs, offset	blt rs, x0, offset	Branch if $<$ zero
bgtz rs, offset	blt x0, rs, offset	Branch if $>$ zero
j offset	jal x0, offset	Jump
jr rs	jalr x0, rs, 0	Jump register
ret	jalr x0, x1, 0	Return from subroutine
tail offset	auipc x6, offset[31:12] jalr x0, x6, offset[11:0]	Tail call far-away subroutine
rdinstret[h] rd	csrrs rd, instret[h], x0	Read instructions-retired counter
rdcycle[h] rd	csrrs rd, cycle[h], x0	Read cycle counter
rdtime[h] rd	csrrs rd, time[h], x0	Read real-time clock
csrr rd, csr	csrrs rd, csr, x0	Read CSR
csrw csr, rs	csrrw x0, csr, rs	Write CSR
csrs csr, rs	csrrs x0, csr, rs	Set bits in CSR
csrc csr, rs	csrrc x0, csr, rs	Clear bits in CSR
csrwi csr, imm	csrrwi x0, csr, imm	Write CSR, immediate
csrsi csr, imm	csrrsi x0, csr, imm	Set bits in CSR, immediate
csrci csr, imm	csrrci x0, csr, imm	Clear bits in CSR, immediate
frcsr rd	csrrs rd, fcsr, x0	Read FP control/status register
fscsr rs	csrrw x0, fcsr, rs	Write FP control/status register
frrm rd	csrrs rd, frm, x0	Read FP rounding mode
fsrm rs	csrrw x0, frm, rs	Write FP rounding mode
frflags rd	csrrs rd, fflags, x0	Read FP exception flags
fsflags rs	csrrw x0, fflags, rs	Write FP exception flags

Figure 3.3: 32 RISC-V pseudoinstructions that rely on x0, the zero register. Appendix A includes includes the RISC-V pseudoinstructions as well as the real instructions. Those that read the 64-bit counters can read by upper 32 bits in RV32I by using the "h" version of the instructions and the lower 32 bits using the plain version. (Tables 20.2 and 20.3 of [Waterman and Asanović 2017] are the basis of this figure.).

Pseudoinstruction	Base Instruction(s)	Meaning
lla rd, symbol	auipc rd, symbol[31:12] addi rd, rd, symbol[11:0]	Load local address
la rd, symbol	*PIC*: auipc rd, GOT[symbol][31:12] l{w|d} rd, rd, GOT[symbol][11:0] *Non-PIC*: Same as lla rd, symbol	Load address
l{b|h|w|d} rd, symbol	auipc rd, symbol[31:12] l{b|h|w|d} rd, symbol[11:0](rd)	Load global
s{b|h|w|d} rd, symbol, rt	auipc rt, symbol[31:12] s{b|h|w|d} rd, symbol[11:0](rt)	Store global
fl{w|d} rd, symbol, rt	auipc rt, symbol[31:12] fl{w|d} rd, symbol[11:0](rt)	Floating-point load global
fs{w|d} rd, symbol, rt	auipc rt, symbol[31:12] fs{w|d} rd, symbol[11:0](rt)	Floating-point store global
li rd, immediate	*Myriad sequences*	Load immediate
mv rd, rs	addi rd, rs, 0	Copy register
not rd, rs	xori rd, rs, -1	One's complement
sext.w rd, rs	addiw rd, rs, 0	Sign extend word
seqz rd, rs	sltiu rd, rs, 1	Set if = zero
fmv.s rd, rs	fsgnj.s rd, rs, rs	Copy single-precision register
fabs.s rd, rs	fsgnjx.s rd, rs, rs	Single-precision absolute value
fneg.s rd, rs	fsgnjn.s rd, rs, rs	Single-precision negate
fmv.d rd, rs	fsgnj.d rd, rs, rs	Copy double-precision register
fabs.d rd, rs	fsgnjx.d rd, rs, rs	Double-precision absolute value
fneg.d rd, rs	fsgnjn.d rd, rs, rs	Double-precision negate
bgt rs, rt, offset	blt rt, rs, offset	Branch if $>$
ble rs, rt, offset	bge rt, rs, offset	Branch if $\leq$
bgtu rs, rt, offset	bltu rt, rs, offset	Branch if $>$, unsigned
bleu rs, rt, offset	bgeu rt, rs, offset	Branch if $\leq$, unsigned
jal offset	jal x1, offset	Jump and link
jalr rs	jalr x1, rs, 0	Jump and link register
call offset	auipc x1, offset[31:12] jalr x1, x1, offset[11:0]	Call far-away subroutine
fence	fence iorw, iorw	Fence on all memory and I/O
fscsr rd, rs	csrrw rd, fcsr, rs	Swap FP control/status register
fsrm rd, rs	csrrw rd, frm, rs	Swap FP rounding mode
fsflags rd, rs	csrrw rd, fflags, rs	Swap FP exception flags

Figure 3.4: 28 RISC-V pseudoinstructions that are independent of x0, the zero register. For la, GOT stands for Global Offset Table, which holds the runtime address of symbols in dynamically linked libraries. Appendix A includes the RISC-V pseudoinstructions as well as the real instructions. (Tables 20.2 and 20.3 of [Waterman and Asanović 2017] are the basis of this figure.)

```
#include <stdio.h>
int main()
{
    printf("Hello, %s\n", "world");
    return 0;
}
```

Figure 3.5: Hello World program in C (hello.c).

```
    .text                       # Directive: enter text section
    .align 2                    # Directive: align code to 2^2 bytes
    .globl main                 # Directive: declare global symbol main
main:                           # label for start of main
    addi sp,sp,-16              # allocate stack frame
    sw   ra,12(sp)              # save return address
    lui  a0,%hi(string1)        # compute address of
    addi a0,a0,%lo(string1)     #    string1
    lui  a1,%hi(string2)        # compute address of
    addi a1,a1,%lo(string2)     #    string2
    call printf                 # call function printf
    lw   ra,12(sp)              # restore return address
    addi sp,sp,16               # deallocate stack frame
    li   a0,0                   # load return value 0
    ret                         # return
    .section .rodata            # Directive: enter read-only data section
    .balign 4                   # Directive: align data section to 4 bytes
string1:                        # label for first string
    .string "Hello, %s!\n"      # Directive: null-terminated string
string2:                        # label for second string
    .string "world"             # Directive: null-terminated string
```

Figure 3.6: Hello World program in RISC-V assembly language (hello.s).

```
00000000 <main>:
   0: ff010113   addi  sp,sp,-16
   4: 00112623   sw    ra,12(sp)
   8: 00000537   lui   a0,0x0
   c: 00050513   mv    a0,a0
  10: 000005b7   lui   a1,0x0
  14: 00058593   mv    a1,a1
  18: 00000097   auipc ra,0x0
  1c: 000080e7   jalr  ra
  20: 00c12083   lw    ra,12(sp)
  24: 01010113   addi  sp,sp,16
  28: 00000513   li    a0,0
  2c: 00008067   ret
```

Figure 3.7: Hello World program in RISC-V machine language (hello.o). The six instructions that are later patched by the linker (locations 8 to 1c) have zero in their address fields. The symbol table included in the object file records the labels and addresses of all the instructions that need to be edited by the linker.

```
000101b0 <main>:
   101b0:  ff010113  addi  sp,sp,-16
   101b4:  00112623  sw    ra,12(sp)
   101b8:  00021537  lui   a0,0x21
   101bc:  a1050513  addi  a0,a0,-1520 # 20a10 <string1>
   101c0:  000215b7  lui   a1,0x21
   101c4:  a1c58593  addi  a1,a1,-1508 # 20a1c <string2>
   101c8:  288000ef  jal   ra,10450 <printf>
   101cc:  00c12083  lw    ra,12(sp)
   101d0:  01010113  addi  sp,sp,16
   101d4:  00000513  li    a0,0
   101d8:  00008067  ret
```

Figure 3.8: Hello World program as RISC-V machine language program after linking. In Unix systems, the file would be named a.out.

Directive	Description
.text	Subsequent items are stored in the text section (machine code).
.data	Subsequent items are stored in the data section (global variables).
.bss	Subsequent items are stored in the bss section (global variables initialized to 0).
.section .foo	Subsequent items are stored in the section named .foo.
.align n	Align the next datum on a 2^n-byte boundary. For example, .align 2 aligns the next value on a word boundary.
.balign n	Align the next datum on a n-byte boundary. For example, .balign 4 aligns the next value on a word boundary.
.globl sym	Declare that label sym is global and may be referenced from other files.
.string "str"	Store the string str in memory and null-terminate it.
.byte b1,..., bn	Store the n 8-bit quantities in successive bytes of memory.
.half w1,...,wn	Store the n 16-bit quantities in successive memory halfwords.
.word w1,...,wn	Store the n 32-bit quantities in successive memory words.
.dword w1,...,wn	Store the n 64-bit quantities in successive memory doublewords.
.float f1,..., fn	Store the n single-precision floating-point numbers in successive memory words.
.double d1,..., dn	Store the n double-precision floating-point numbers in successive memory doublewords.
.option rvc	Compress subsequent instructions (see Chapter 7).
.option norvc	Don't compress subsequent instructions.
.option relax	Allow linker relaxations for subsequent instructions.
.option norelax	Don't allow linker relaxations for subsequent instructions.
.option pic	Subsequent instructions are position-independent code.
.option nopic	Subsequent instructions are position-dependent code.
.option push	Push the current setting of all .options to a stack, so that a subsequent .option pop will restore their value.
.option pop	Pop the option stack, restoring all .options to their setting at the time of the last .option push.

Figure 3.9: Common RISC-V assembler directives.

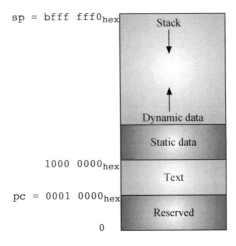

sp = bfff fff0$_{hex}$

1000 0000$_{hex}$

pc = 0001 0000$_{hex}$

0

Figure 3.10: RV32I allocation of memory to program and data. The high addresses are the top of the figure and the low addresses are the bottom. In this RISC-V software convention, the stack pointer (sp) starts at bfff fff0$_{hex}$ and grows down toward the Static data. The *text* (program code) starts at 0001 0000$_{hex}$ and includes the statically-linked libraries. The Static data starts immediately above the text region; in this example, we assume that address is 1000 0000$_{hex}$. Dynamic data, allocated in C by malloc(), is just above the Static data. Called the heap, it grows upward toward the stack. It includes the dynamically-linked libraries.

3.4   Linker

Rather than compile all the source code every time one file changes, the linker allows individual files to be compiled and assembled separately. It "stitches" the new object code together with existing machine language modules, such as libraries. It derives its name from one of its tasks, which is to all edit the links of the jump and link instructions in the object file. In fact, linker is short for link editor, which was the historical name of this step in Figure 3.1. In Unix systems, the input to the linker are files with the .o suffix (e.g., foo.o, libc.o), and its output is the a.out file. For MS-DOS, the inputs are files with the suffix .OBJ or .LIB and the output is a .EXE file.

Figure 3.10 shows the addresses of the regions of memory allocated for code and data in a typical RISC-V program. The linker must adjust the program and data addresses of instructions in all the object files to match addresses in this figure. It is less work for the linker if the input files are *position independent code* (*PIC*). PIC means that all the branches to instructions and references to data inside the file are correct wherever the code is placed. As mentioned in Chapter 2, the PC-relative branch of RV32I makes PIC much easier to fulfill.

In addition to the instructions, each object file contains a symbol table that includes all the labels in the program that must be given addresses as part of the linking process. This list includes labels to data as well as to code. Figure 3.6 has two data labels to be set (string1 and string2) and two code labels to be assigned in (main and printf). A 32-bit address is hard to fit in one 32-bit instruction, so the linker must adjust two instructions per label in the RV32I code, as Figure 3.6 shows: lui and addi for data addresses, and auipc and jalr for code addresses. Figure 3.8 is the final linked a.out version of the object file in Figure 3.7.

RISC-V compilers support several ABIs, depending on whether the F and D extensions are present. For RV32, the ABIs are named ilp32, ilp32f, and ilp32d. ilp32 means that the C language data types int, long, and pointer are all 32 bits; the optional suffix indicates how floating-point arguments are passed. In ilp32, floating-point arguments are passed in integer registers. In ilp32f, single-precision floating-point arguments are passed in floating-point registers. In ilp32d, double-precision floating-point arguments are also passed in floating-point registers.

Naturally, to pass a floating-point argument in a floating-point register, you need the corresponding floating-point ISA extension F or D (see Chapter 5). So, to compile code for RV32I (GCC flag '-march=rv32i'), you must use the ilp32 ABI (GCC flag '-mabi=ilp32'). On the other hand, having floating-point instructions doesn't mean the calling convention is required to use them; so, for example, RV32IFD is compatible with all three ABIs: ilp32, ilp32f, and ilp32d.

The linker checks that the program's ABI matches all of its libraries. Although the compiler supports many combinations of ISA extensions and ABIs, only a few sets of libraries might be installed. Hence, a common pitfall is attempting to link a program without having compatible libraries installed. The linker will not produce a helpful diagnostic message in this case; it will simply attempt to link with an incompatible library, then inform you of the incompatibility. This pitfall generally occurs only when compiling on one computer for a different computer (*cross compiling*).

■ *Elaboration: Linker relaxation*

The jump and link instruction has a 20-bit PC-relative address field, so a single instruction can jump far. While the compiler produces two instructions for each external function, quite often only one instruction is necessary. Since this optimization saves both time and space, linkers will make passes over the code to replace two instructions with one whenever it can. Because a pass might shrink distance between a call and the function so that it now fits in a single instruction, the linker keeps optimizing the code until there are no further changes. This process is called *Linker relaxation*, with the name referring to relaxation techniques for solving systems of equations. In addition to procedure calls, the RISC-V linker relaxes data addressing to use the global pointer when the datum lies within ± 2 KiB of gp, removing a lui or auipc. It similarly relaxes thread-local storage addressing when the datum lies within ± 2 KiB of tp.

3.5 Static vs. Dynamic Linking

The prior section describes *static linking*, where all potential library code is linked and then loaded together before execution. Such libraries can be relatively large, so linking a popular library into multiple programs wastes memory. Moreover, the libraries are bound when linked—even when they are updated later to fix bugs—forcing the statically-linked code to use the old, buggy version.

To avoid both problems, most systems rely on *dynamic linking*, where the desired external function is loaded and linked to the program only after it is first called; if never called, it's never loaded and linked. Every call after the first uses a fast link, so the dynamic overhead is only paid once. When a program starts it links in the current version of the library functions it needs, which is how it gets the newest version. Moreover, if multiple programs use the same dynamically linked library, the code in the library appears only once in memory.

Architects typically measure processor performance using benchmarks that are statically linked despite most real programs having dynamic links. The excuse is that users interested in performance should link statically, but it's a poor justification. It makes more sense to accelerate performance of real programs, not benchmarks.

The code that the compiler generates resembles that for static linking. Instead of jumping to the real function, it jumps to a short (three-instruction) stub function. The stub function loads the address of the real function from a table in memory, then jumps to it. However, on the first call, the table lacks the address of the real function, but instead contains the address of the dynamic-linking routine. When invoked, the dynamic linker uses the symbol table to find the real function, copies it into memory, and then updates the table to point to the real function. Each subsequent call pays only the three-instruction overhead of the stub function.

3.6   Loader

A program like the one in Figure 3.8 is an executable file kept in the computer's storage. When one is to be run, the loader's job is to load it into memory and jump to the starting address. The "loader" today is the operating system; stated alternatively, loading a.out is one of many tasks of an operating system.

Loading is a little trickier for dynamically-linked programs. Instead of simply starting the program, the operating system starts the dynamic linker. It in turn starts the desired program, and then handles all first-time external calls, copies the functions into memory, and edits the program after each call to point it to the correct function.

3.7   Concluding Remarks

Keep it simple, stupid.

—Kelly Johnson, aeronautical engineer who coined the "KISS Principle," 1960

Programmability

Cost

Performance

Elegance

The assembler enhances the simple RISC-V ISA with 60 pseudoinstructions that make RISC-V code easier to read and to write without increasing hardware costs. Simply dedicating one RISC-V register to zero enables many of these helpful operations. The Load Upper Immediate (lui) and Add Upper Immediate to PC (auipc) instructions make it easier for the compiler and linker to adjust addresses for external data and functions, and PC-relative branching makes it easier to help the linker with position-independent code. Having plenty of registers enables a calling convention that makes function call and return faster by reducing the number of register spills and restores.

RISC-V offers a tasteful collection of simple but impactful mechanisms that reduce cost, improve performance, and make it easier to program.

3.8   To Learn More

D. A. Patterson and J. L. Hennessy. *Computer Organization and Design RISC-V Edition: The Hardware Software Interface*. Morgan Kaufmann, 2017.

TIS Committee. Tool interface standard (TIS) executable and linking format (ELF) specification version 1.2. *TIS Committee*, 1995.

A. Waterman and K. Asanović, editors. *The RISC-V Instruction Set Manual, Volume I: User-Level ISA, Version 2.2*. May 2017. URL https://riscv.org/specifications/.

Notes

[1]http://parlab.eecs.berkeley.edu

4 RV32M: Multiply and Divide

Entities should not be multiplied beyond necessity.

—William of Occam, 1320

4.1 Introduction

RV32M adds integer multiply and divide instructions to RV32I. Figure 4.1 is a graphical representation of the RV32M extension instruction set and Figure 4.2 lists their opcodes.

Divide is straightforward. Recall that

$$Quotient = (Dividend - Remainder) \div Divisor$$

or alternatively

$$Dividend = Quotient \times Divisor + Remainder$$

$$Remainder = Dividend - (Quotient \times Divisor)$$

RV32M has divide instructions for both signed and unsigned integers: divide (`div`) and divide unsigned (`divu`), which place the quotient into the destination register. Less frequently, programmers want the remainder instead of the quotient, so RV32M offers remainder (`rem`) and remainder unsigned (`remu`), which write the remainder instead of the quotient.

RV32M

Figure 4.1: Diagram of the RV32M instructions.

31	25	24	20	19	15	14	12	11	7	6	0	
0000001		rs2		rs1		000		rd		0110011		R mul
0000001		rs2		rs1		001		rd		0110011		R mulh
0000001		rs2		rs1		010		rd		0110011		R mulhsu
0000001		rs2		rs1		011		rd		0110011		R mulhu
0000001		rs2		rs1		100		rd		0110011		R div
0000001		rs2		rs1		101		rd		0110011		R divu
0000001		rs2		rs1		110		rd		0110011		R rem
0000001		rs2		rs1		111		rd		0110011		R remu

Figure 4.2: RV32M opcode map has instruction layout, opcodes, format type, and names. (Table 19.2 of [Waterman and Asanović 2017] is the basis of this figure.)

```
# Compute unsigned division of a0 by 3 using multiplication.
0: aaaab2b7    lui   t0,0xaaaab  # t0 = 0xaaaaaaab
4: aab28293    addi  t0,t0,-1365 #    = ~ 2^32 / 1.5
8: 025535b3    mulhu a1,a0,t0    # a1 = ~ (a0 / 1.5)
c: 0015d593    srli  a1,a1,0x1   # a1 = (a0 / 3)
```

Figure 4.3: RV32M code to divide by a constant by multiplying. It takes careful numerical analysis to show that this algorithm works for any dividend, and for some other divisors, the correction step is more complicated. The proof of correctness, and the algorithm for generating the reciprocals and correction steps, is in [Granlund and Montgomery 1994].

The multiply equation is simply:

$$Product = Multiplicand \times Multiplier$$

It's more complicated than divide because the size of the product is the sum of the sizes of the multiplier and the multiplicand; multiplying two 32-bit numbers yields a 64-bit product. To produce a properly signed or unsigned 64-bit product, RISC-V has four multiply instructions. To get the integer 32-bit product—the lower 32-bits of the full product—use `mul`. To get the upper 32 bits of the 64-bit product, use `mulh` if both operands are signed, `mulhu` if both operands are unsigned, and `mulhsu` is one is signed, and the other is unsigned. Since it would complicate the hardware to write the 64-bit product into two 32-bit registers in one instruction, RV32M requires two multiply instructions to produce the 64-bit product.

For many microprocessors, integer division is a relatively slow operation. As mentioned above, right shifts can replace unsigned division by powers of 2. It turns out that division by other constants can be optimized, too, by multiplying by the approximate reciprocal then applying a correction to the upper half of the product. For example, Figure 4.3 shows the code for unsigned division by 3.

What's Different? ARM-32 long had multiply but no divide instruction. Divide didn't become mandatory until 2005, almost 20 years after the first ARM processor. MIPS-32 uses special registers (HI and LO) as the sole destination registers for multiply and divide instructions. While this design reduced the complexity of early MIPS implementations, it takes an extra move instruction to use the result of the multiply or divide, potentially reducing performance. The HI and LO registers also increase the architectural state, making it slightly slower to switch between tasks.

`sll` **can do signed and unsigned multiplication by** 2^i. For example, if a2 = 16 (2^4) then `slli t2,a1,4` produces the same value as `mul t2,a1,a2`.

Performance

For almost all processors, multiplies are slower than shifts or adds and divides are much slower than multiplies.

■ *Elaboration:* `mulh` *and* `mulhu` *can check for overflow in multiplication.*

There is no overflow when using `mul` for unsigned multiplication if the result of `mulhu` is zero. Similarly, there is no overflow when using `mul` for signed multiplication if all bits in the result of `mulh` match the sign bit of the result of `mul`, i.e., equals 0 if positive or ffff ffff$_{hex}$ if negative.

■ *Elaboration: It's also easy to check for divide by zero.*

Just add a `beqz` test of the dividend before the divide. RV32I doesn't trap on divide by zero because few programs want that behavior, and the ones that do can easily check for zero in software. Of course, divides by constants never need checks.

■ *Elaboration:* `mulhsu` *is useful for multi-word signed multiplication.*

`mulhsu` generates the upper half of the product when the multiplier is signed and the multiplicand is unsigned. It is as a substep of multi-word signed multiplication when multiplying the most-significant word of the multiplier (which contains the sign bit) with the less-significant words of the multiplicand (which are unsigned). This instruction improves performance of multi-word multiplication by about 15%.

4.2   Concluding Remarks

The cheapest, fastest, and most reliable components are those that aren't there.

—C. Gordon Bell, architect of prominent minicomputers

Cost

To offer the smallest RISC-V processor for embedded applications, multiply and divide are part of the first optional standard extension of RISC-V. Nevertheless, many RISC-V processors will include RV32M.

4.3   To Learn More

T. Granlund and P. L. Montgomery. Division by invariant integers using multiplication. In *ACM SIGPLAN Notices*, volume 29, pages 61–72. ACM, 1994.

A. Waterman and K. Asanović, editors. *The RISC-V Instruction Set Manual, Volume I: User-Level ISA, Version 2.2.* May 2017. URL https://riscv.org/specifications/.

Notes

[1] http://parlab.eecs.berkeley.edu

5 RV32F and RV32D: Single- and Double-Precision Floating Point

Perfection is finally attained not when there is no longer anything to add, but when there is no longer anything to take away.

—Antoine de Saint Exup'ery, L'Avion, 1940

5.1 Introduction

Although RV32F and RV32D are separate, optional instruction set extensions, they are often included together. Given single- and double-precision (32- and 64-bit) versions of nearly all floating-point instructions, for brevity we present them in one chapter. Figure 5.1 is a graphical representation of the RV32F and RV32D extension instruction sets. Figure 5.2 lists the opcodes of RV32F and Figure 5.3 lists the opcodes of RV32D. Like virtually all other modern ISAs, RISC-V obeys the IEEE 754-2008 floating-point standard [IEEE Standards Committee 2008].

5.2 Floating-Point Registers

Performance

RV32F and RV32D use 32 separate f registers instead of the x registers. The main reason for the two sets of registers is that processors can improve performance by doubling the register capacity and bandwidth by having two sets of registers without increasing the space for the register specifier in the cramped RISC-V instruction format. The major impact on the instruction set is to have new instructions to load and store the f registers and to transfer data between the x and f registers. Figure 5.4 lists the RV32D and RV32F registers and their names as determined by the RISC-V ABI.

If a processor has both RV32F and RV32D, the single-precision data uses only the lower 32 bits of the f registers. Unlike x0 in RV32I, register f0 is *not* hardwired to 0 but is an alterable register like all the other 31 f registers.

The IEEE 754-2008 standard provides several ways to round floating-point arithmetic, which are helpful to determine error bounds and in writing numerical libraries. The most accurate and most common is round to nearest even (RNE). The rounding mode is set in the floating-point control and status register fcsr. Figure 5.5 shows fcsr and lists the rounding options. It also holds the accrued exception flags that the standard requires.

What's Different? Both ARM-32 and MIPS-32 have 32 single-precision floating-point registers but only 16 double-precision registers. They both map two single-precision registers

RV32F and RV32D

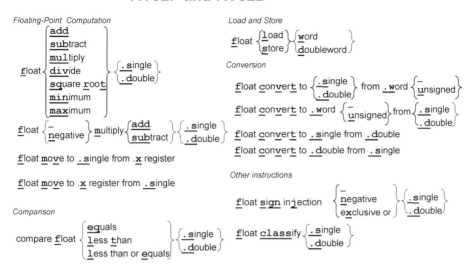

Figure 5.1: Diagram of the RV32F and RV32D instructions.

into the left and right 32-bit halves of a double-precision register. x86-32 floating-point arithmetic didn't have any registers, but used a stack instead. The stack entries were 80-bits wide to improve accuracy, so loads covert 32-bit or 64-bit operands to 80 bits, and vice versa for stores. A subsequent version of x86-32 added 8 traditional 64-bit floating-point registers and associated instructions. Unlike RV32FD and MIPS-32, ARM-32 and x86-32 overlooked instructions to move data directly between floating-point and integer registers. The only solution is to store a floating-point register in memory and then load it from memory to an integer register, and vice versa.

> **Having only 16 double-precision registers was the most painful ISA error in MIPS** according to John Mashey, one of its architects.

■ *Elaboration: RV32FD allows the rounding mode to be set per instruction.*

Called *static rounding*, it helps performance when you only need to change the rounding mode for one instruction. The default is to use the dynamic rounding mode in the `fcsr`. Static rounding is specified as an optional last argument, as `fadd.s ft0, ft1, ft2, rtz` will round towards zero, irrespective of `fcsr`. The caption of Figure 5.5 lists the names of the rounding modes.

5.3 Floating-Point Loads, Stores, and Arithmetic

RISC-V has two load instructions (`flw`, `fld`) and two store instructions (`fsw`, `fsd`) for RV32F and RV32D. The have the same addressing mode and instruction format as `lw` and `sw`.

Adding to the standard arithmetic operations (`fadd.s`, `fadd.d`, `fsub.s`, `fsub.d`, `fmul.s`, `fmul.d`, `fdiv.s`, `fdiv.d`), RV32F and RV32D include square root (`fsqrt.s`, `fsqrt.d`). They also have minimum and maximum (`fmin.s`, `fmin.d`, `fmax.s`, `fmax.d`), which write the smaller or larger values from the pair of source operands without using a branch instruction.

> **Unlike integer arithmetic, the size of the product from a floating-point multiply is the same as its operands.** Also, RV32F and RF32D omit floating-point remainder instructions.

31 27	26 25 24	20	19 15	14 12	11 7	6 0		
imm[11:0]			rs1	010	rd	0000111	I	flw
imm[11:5]		rs2	rs1	010	imm[4:0]	0100111	S	fsw
rs3	00	rs2	rs1	rm	rd	1000011	R4	fmadd.s
rs3	00	rs2	rs1	rm	rd	1000111	R4	fmsub.s
rs3	00	rs2	rs1	rm	rd	1001011	R4	fnmsub.s
rs3	00	rs2	rs1	rm	rd	1001111	R4	fnmadd.s
0000000		rs2	rs1	rm	rd	1010011	R	fadd.s
0000100		rs2	rs1	rm	rd	1010011	R	fsub.s
0001000		rs2	rs1	rm	rd	1010011	R	fmul.s
0001100		rs2	rs1	rm	rd	1010011	R	fdiv.s
0101100		00000	rs1	rm	rd	1010011	R	fsqrt.s
0010000		rs2	rs1	000	rd	1010011	R	fsgnj.s
0010000		rs2	rs1	001	rd	1010011	R	fsgnjn.s
0010000		rs2	rs1	010	rd	1010011	R	fsgnjx.s
0010100		rs2	rs1	000	rd	1010011	R	fmin.s
0010100		rs2	rs1	001	rd	1010011	R	fmax.s
1100000		00000	rs1	rm	rd	1010011	R	fcvt.w.s
1100000		00001	rs1	rm	rd	1010011	R	fcvt.wu.s
1110000		00000	rs1	000	rd	1010011	R	fmv.x.w
1010000		rs2	rs1	010	rd	1010011	R	feq.s
1010000		rs2	rs1	001	rd	1010011	R	flt.s
1010000		rs2	rs1	000	rd	1010011	R	fle.s
1110000		00000	rs1	001	rd	1010011	R	fclass.s
1101000		00000	rs1	rm	rd	1010011	R	fcvt.s.w
1101000		00001	rs1	rm	rd	1010011	R	fcvt.s.wu
1111000		00000	rs1	000	rd	1010011	R	fmv.w.x

Figure 5.2: **RV32F opcode map has instruction layout, opcodes, format type, and names. The primary difference in the encodings between this and the next figure is bit 12 is a 0 for the first two instructions and bit 25 is a 0 for the rest of the instructions where both bits are 1 in RV32D. (Table 19.2 of [Waterman and Asanović 2017] is the basis of this figure.)**

31	27	26 25	24	20	19	15	14	12	11	7	6	0		
imm[11:0]					rs1		011		rd		0000111		I	fld
imm[11:5]			rs2		rs1		011		imm[4:0]		0100111		S	fsd
rs3		01	rs2		rs1		rm		rd		1000011		R4	fmadd.d
rs3		01	rs2		rs1		rm		rd		1000111		R4	fmsub.d
rs3		01	rs2		rs1		rm		rd		1001011		R4	fnmsub.d
rs3		01	rs2		rs1		rm		rd		1001111		R4	fnmadd.d
0000001			rs2		rs1		rm		rd		1010011		R	fadd.d
0000101			rs2		rs1		rm		rd		1010011		R	fsub.d
0001001			rs2		rs1		rm		rd		1010011		R	fmul.d
0001101			rs2		rs1		rm		rd		1010011		R	fdiv.d
0101101			00000		rs1		rm		rd		1010011		R	fsqrt.d
0010001			rs2		rs1		000		rd		1010011		R	fsgnj.d
0010001			rs2		rs1		001		rd		1010011		R	fsgnjn.d
0010001			rs2		rs1		010		rd		1010011		R	fsgnjx.d
0010101			rs2		rs1		000		rd		1010011		R	fmin.d
0010101			rs2		rs1		001		rd		1010011		R	fmax.d
0100000			00001		rs1		rm		rd		1010011		R	fcvt.s.d
0100001			00000		rs1		rm		rd		1010011		R	fcvt.d.s
1010001			rs2		rs1		010		rd		1010011		R	feq.d
1010001			rs2		rs1		001		rd		1010011		R	flt.d
1010001			rs2		rs1		000		rd		1010011		R	fle.d
1110001			00000		rs1		001		rd		1010011		R	fclass.d
1100001			00000		rs1		rm		rd		1010011		R	fcvt.w.d
1100001			00001		rs1		rm		rd		1010011		R	fcvt.wu.d
1101001			00000		rs1		rm		rd		1010011		R	fcvt.d.w
1101001			00001		rs1		rm		rd		1010011		R	fcvt.d.wu

Figure 5.3: **RV32D opcode map has instruction layout, opcodes, format type, and names. There are some instructions in these two figures do not simply differ by data width. This figure uniquely has** fcvt.s.d **and** fcvt.d.s **while the other has** fmv.x.w **and** fmv.w.x**. (Table 19.2 of [Waterman and Asanović 2017] is the basis of this figure.)**

63	32	31	0	
		f0 / ft0		FP Temporary
		f1 / ft1		FP Temporary
		f2 / ft2		FP Temporary
		f3 / ft3		FP Temporary
		f4 / ft4		FP Temporary
		f5 / ft5		FP Temporary
		f6 / ft6		FP Temporary
		f7 / ft7		FP Temporary
		f8 / fs0		FP Saved register
		f9 / fs1		FP Saved register
		f10 / fa0		FP Function argument, return value
		f11 / fa1		FP Function argument, return value
		f12 / fa2		FP Function argument
		f13 / fa3		FP Function argument
		f14 / fa4		FP Function argument
		f15 / fa5		FP Function argument
		f16 / fa6		FP Function argument
		f17 / fa7		FP Function argument
		f18 / fs2		FP Saved register
		f19 / fs3		FP Saved register
		f20 / fs4		FP Saved register
		f21 / fs5		FP Saved register
		f22 / fs6		FP Saved register
		f23 / fs7		FP Saved register
		f24 / fs8		FP Saved register
		f25 / fs9		FP Saved register
		f26 / fs10		FP Saved register
		f27 / fs11		FP Saved register
		f28 / ft8		FP Temporary
		f29 / ft9		FP Temporary
		f30 / ft10		FP Temporary
		f31 / ft11		FP Temporary
32		32		

Figure 5.4: The floating-point registers of RV32F and RV32D. The single-precision registers occupy the rightmost half of the 32 double-precision registers. Chapter 3 explains the RISC-V calling convention for the floating-point registers, the rationale behind the FP Argument registers (fa0-fa7), FP Saved registers (fs0-fs11), and FP Temporaries (ft0-ft11). (Table 20.1 of [Waterman and Asanović 2017] is the basis of this figure.)

31	8	7	5	4	3	2	1	0
Reserved		Rounding Mode (frm)		Accrued Exceptions (fflags)				
				NV	DZ	OF	UF	NX
24		3		1	1	1	1	1

Figure 5.5: Floating-point control and status register. It holds the rounding modes and the exception flags. The rounding modes are round to nearest, ties to even (rte, 000 in frm); round towards zero (rtz, 001); round down, towards $-\infty$ (rdn, 010); round up, towards $+\infty$ (rup, 011); and round to nearest, ties to max magnitude (rmm, 100). The five accrued exception flags indicate the exception conditions that have arisen on any floating-point arithmetic instruction since the field was last reset by software: NV is Invalid Operation; DZ is Divide by Zero; OF is Overflow; UF is Underflow; and NX is Inexact. (Figure 8.2 of [Waterman and Asanović 2017] is the basis of this figure.)

Many floating-point algorithms, such as matrix multiply, perform a multiply immediately followed by an addition or a subtraction. Hence, RISC-V offers instructions that multiply two operands and then either add (fmadd.s, fmadd.d) or subtract (fmsub.s, fmsub.d) a third operand to that product before writing the sum. It also has versions that negate the product before adding or subtracting the third operand: fnmadd.s, fnmadd.d, fnmsub.s, fnmsub.d. These *fused* multiply-add instructions are required by the IEEE 754-2008 standard for their increased accuracy: they round only once (after the add) rather than twice (after the multiply, then after the add). Skipping the intermediate rounding makes a big difference when the product and addend have similar magnitudes but opposite signs, which causes most of the mantissa bits to cancel in subtraction. These instructions need a new instruction format to specify 4 registers, called R4. Figures 5.2 and 5.3 show the R4 format, which is a variation of the R format.

Performance

Instead of floating-point branch instructions, RV32F and RV32D supply comparison instructions that set an integer register to 1 or 0 based on comparison of two floating-point registers: feq.s, feq.d, flt.s, flt.d, fle.s, fle.d. These instructions allow an integer branch instruction to jump based on a floating-point condition. For example, this code branches to Exit if f1 < f2:

```
flt x5, f1, f2   # x5 = 1 if f1 < f2; otherwise x5 = 0
bne x5, x0, Exit # if x5 != 0, jump to Exit
```

5.4   Floating-Point Converts and Moves

RV32F and RV32D have instructions that perform all combinations of useful conversions between 32-bit signed integers, 32-bit unsigned integers, 32-bit floating point, and 64-bit floating point. Figure 5.6 displays these 10 instructions by source data type and converted destination data type.

RV32F also offers instructions to move data to x from f registers (fmv.x.w) and vice versa (fmv.w.x).

5.5   Miscellaneous Floating-Point Instructions

RV32F and RV32D offer unusual instructions that help with math libraries as well as provide useful pseudoinstructions. (The IEEE 754 floating-point standard requires a way to copy and manipulate signs and to classify floating-point data, which inspired these instructions.)

To	From			
	32b signed integer (w)	32b unsigned integer (wu)	32b floating point (s)	64b floating point (d)
32b signed integer (w)	–	–	fcvt.w.s	fcvt.w.d
32b unsigned integer (wu)	–	–	fcvt.wu.s	fcvt.wu.d
32b floating point (s)	fcvt.s.w	fcvt.s.wu	–	fcvt.s.d
64b floating point (d)	fcvt.d.w	fcvt.d.wu	fcvt.d.s	–

Figure 5.6: RV32F and RV32D conversion instructions. The columns list the source data types and the rows show the converted destination data type.

The first is the *sign-injection* instructions, which copy everything from the first source register but the sign bit. The value of the sign bit depends on the instruction:

1. Float sign inject (fsgnj.s, fsgnj.d): the result's sign bit is rs2's sign bit.

2. Float sign inject negative (fsgnjn.s, fsgnjn.d): the result's sign bit is the opposite of rs2's sign bit.

3. Float sign inject exclusive-or (fsgnjx.s, fsgnjx.d): the sign bit is the XOR of the sign bits of rs1 and rs2.

As well as helping with sign manipulation in math libraries, sign-injection instructions provide three popular floating-point pseudoinstructions (see Figure 3.4 on page 37):

1. Copy floating-point register:
   fmv.s rd,rs is really fsgnj.s rd,rs,rs and
   fmv.d rd,rs is really fsgnj.d rd,rs,rs.

2. Negate:
   fneg.s rd,rs maps to fsgnjn.s rd,rs,rs and
   fneg.d rd,rs maps to fsgnjn.d rd,rs,rs.

3. Absolute value (since $0 \oplus 0 = 0$ and $1 \oplus 1 = 0$):
   fabs.s rd,rs becomes fsgnjx.s rd,rs,rs and
   fabs.d rd,rs becomes fsgnjx.d rd,rs,rs.

The second unusual floating-point instruction is classify (fclass.s, fclass.d). Classify instructions are also a great aid to math libraries. They test a source operand to see which of 10 floating-point properties apply (see the table below), and then write a mask into the lower 10 bits of the destination integer register with the answer. Only one of the ten bits is set to 1, with the rest set to 0s.

```
void daxpy(size_t n, double a, const double x[], double y[])
{
  for (size_t i = 0; i < n; i++) {
    y[i] = a*x[i] + y[i];
  }
}
```

Figure 5.7: The floating-point intensive DAXPY program in C.

ISA	ARM-32	ARM Thumb-2	MIPS-32	microMIPS	x86-32	RV32FD	RV32FD+RV32C
Instructions	10	10	12	12	16	11	11
Per Loop	6	6	7	7	6	7	7
Bytes	40	28	48	32	50	44	28

Figure 5.8: Number of instructions and code size of DAXPY for four ISAs. It lists number of instructions per loop and total. Chapter 7 describes ARM Thumb-2, microMIPS, and RV32C.

$x[rd]$ bit	Meaning
0	f[$rs1$] is $-\infty$.
1	f[$rs1$] is a negative normal number.
2	f[$rs1$] is a negative subnormal number.
3	f[$rs1$] is -0.
4	f[$rs1$] is $+0$.
5	f[$rs1$] is a positive subnormal number.
6	f[$rs1$] is a positive normal number.
7	f[$rs1$] is $+\infty$.
8	f[$rs1$] is a signaling NaN.
9	f[$rs1$] is a quiet NaN.

5.6 Comparing RV32FD, ARM-32, MIPS-32, and x86-32 using DAXPY

We'll now do a head-to-head comparison using DAXPY as our floating-point benchmark (Figure 5.7). It calculates $Y = a \times X + Y$ in double-precision, where X and Y are vectors and a is a scalar. Figure 5.8 summarizes the number of instructions and number of bytes in DAXPY of programs for the four ISAs. Their code is in Figures 5.9 to 5.12.

As was the case for Insertion Sort in Chapter 2, despite its emphasis on simplicity, the RISC-V version again has about the same or fewer instructions, and the code sizes of the architectures are quite close. In this example, the compare-and-execute branches of RISC-V save as many instructions as do the fancier address modes and the push and pop instructions of ARM-32 and x86-32.

5.7 Concluding Remarks

Less is More.

—Robert Browning, 1855. The Minimalist school of (building) architecture adopted this poem as an axiom in the 1980s.

The name DAXPY come from the formula itself: Double-precision A times X Plus Y. The single-precision version is called SAXPY.

Simplicity

Performance

The IEEE 754-2008 floating-point standard [IEEE Standards Committee 2008] defines the floating-point data types, the accuracy of computation, and the required operations. Its success greatly reduces the difficulty of porting floating-point programs, and it also means that the floating-point ISAs are probably more uniform than are the equivalent in other chapters.

■ *Elaboration: 16-bit, 128-bit, and decimal floating-point arithmetic*

The revised IEEE floating-point standard (IEEE 754-2008) describes several new formats beyond single- and double-precision, which they call *binary32* and *binary64*. The least surprising addition is quadruple precision, named *binary128*. RISC-V has a tentative extension planned for it called RV32Q (see Chapter 11). The standard also provided two more sizes for binary data interchange, indicating that programmers might store these numbers in memory or storage but shouldn't expect to be able to compute in these sizes. They are half-precision (*binary16*) and octuple precision (*binary256*). Despite the standard's intent, GPUs do compute in half-precision as well as keep them in memory. The plan for RISC-V is to include half-precision in the vector instructions (RV32V in Chapter 8), with the proviso that processors supporting vector half-precision will also add half-precision scalar instructions. The surprising addition to the revised standard is decimal floating point, for which RISC-V has set aside RV32L (see Chapter 11). The three self-explanatory decimal formats are called *decimal32*, *decimal64*, and *decimal128*.

5.8 To Learn More

IEEE Standards Committee. 754-2008 IEEE standard for floating-point arithmetic. *IEEE Computer Society Std*, 2008.

A. Waterman and K. Asanović, editors. *The RISC-V Instruction Set Manual, Volume I: User-Level ISA, Version 2.2*. May 2017. URL https://riscv.org/specifications/.

Notes

[1] http://parlab.eecs.berkeley.edu

```
# RV32FD (7 insns in loop; 11 insns/44 bytes total; 28 bytes RVC)
# a0 is n, a1 is pointer to x[0], a2 is pointer to y[0], fa0 is a
   0: 02050463 beqz    a0,28              # if n == 0, jump to Exit
   4: 00351513 slli    a0,a0,0x3          # a0 = n*8
   8: 00a60533 add     a0,a2,a0           # a0 = address of x[n] (last element)
Loop:
   c: 0005b787 fld     fa5,0(a1)          # fa5 = x[]
  10: 00063707 fld     fa4,0(a2)          # fa4 = y[]
  14: 00860613 addi    a2,a2,8            # a2++ (increment pointer to y)
  18: 00858593 addi    a1,a1,8            # a1++ (increment pointer to x)
  1c: 72a7f7c3 fmadd.d fa5,fa5,fa0,fa4    # fa5 = a*x[i] + y[i]
  20: fef63c27 fsd     fa5,-8(a2)         # y[i] = a*x[i] + y[i]
  24: fea614e3 bne     a2,a0,c            # if i != n, jump to Loop
Exit:
  28: 00008067         ret                # return
```

Figure 5.9: RV32D code for DAXPY in Figure 5.7. The address in hexadecimal is on the left, the machine language code in hexadecimal is next, and then the assembly language instruction followed by a comment. The compare-and-branch instructions avoid the two compare instructions in the code of ARM-32 and x86-32.

```
# ARM-32 (6 insns in loop; 10 insns/40 bytes total; 28 bytes Thumb-2)
# r0 is n, d0 is a, r1 is pointer to x[0], r2 is pointer to y[0]
   0: e3500000 cmp      r0, #0             # compare n to 0
   4: 0a000006 beq      24 <daxpy+0x24>    # if n == 0, jump to Exit
   8: e0820180 add      r0, r2, r0, lsl #3 # r0 = address of x[n] (last element)
Loop:
   c: ecb16b02 vldmia   r1!,{d6}           # d6 = x[i], increment pointer to x
  10: ed927b00 vldr     d7, [r2]           # d7 = y[i]
  14: ee067b00 vmla.f64 d7, d6, d0         # d7 = a*x[i] + y[i]
  18: eca27b02 vstmia   r2!, {d7}          # y[i] = a*x[i] + y[i], incr. ptr to y
  1c: e1520000 cmp      r2, r0             # i vs. n
  20: 1afffff9 bne      c <daxpy+0xc>      # if i != n, jump to Loop
Exit:
  24: e12fff1e bx       lr                 # return
```

Figure 5.10: ARM-32 code for DAXPY in Figure 5.7. The autoincrement addressing mode of ARM-32 saves two instructions as compared to RISC-V. Unlike Insertion Sort, there is no need to push and pop registers for DAXPY in ARM-32.

```
# MIPS-32 (7 insns in loop; 12 insns/48 bytes total; 32 bytes microMIPS)
# a0 is n, a1 is pointer to x[0], a2 is pointer to y[0], f12 is a
   0: 10800009 beqz    a0,28 <daxpy+0x28> # if n == 0, jump to Exit
   4: 000420c0 sll     a0,a0,0x3          # a0 = n*8 (filled branch delay slot)
   8: 00c42021 addu    a0,a2,a0           # a0 = address of x[n] (last element)
Loop:
   c: 24c60008 addiu   a2,a2,8            # a2++ (increment pointer to y)
  10: d4a00000 ldc1    $f0,0(a1)          # f0 = x[i]
  14: 24a50008 addiu   a1,a1,8            # a1++ (increment pointer to x)
  18: d4c2fff8 ldc1    $f2,-8(a2)         # f2 = y[i]
  1c: 4c406021 madd.d  $f0,$f2,$f12,$f0   # f0 = a*x[i] + y[i]
  20: 14c4fffa bne     a2,a0,c <daxpy+0xc> # if i != n, jump to Loop
  24: f4c0fff8 sdc1    $f0,-8(a2)         # y[i] = a*x[i] + y[i] (filled delay slot)
Exit:
  28: 03e00008 jr      ra                 # return
  2c: 00000000 nop                        # (unfilled branch delay slot)
```

Figure 5.11: MIPS-32 code for DAXPY in Figure 5.7. Two of the three branch delay slots are filled with useful instructions. The ability to check for equality between two registers avoids the two compare instructions found in ARM-32 and x86-32. Unlike integer loads, floating-point loads have no delay slot.

```
# x86-32 (6 insns in loop; 16 insns/50 bytes total)
# eax is i, n is in memory at esp+0x8, a is in memory at esp+0xc
# pointer to x[0] is in memory at esp+0x14
# pointer to y[0] is in memory at esp+0x18
   0: 53                 push   ebx                  # save ebx
   1: 8b 4c 24 08        mov    ecx,[esp+0x8]        # ecx has copy of n
   5: c5 fb 10 4c 24 0c  vmovsd xmm1,[esp+0xc]       # xmm1 has a copy of a
   b: 8b 5c 24 14        mov    ebx,[esp+0x14]       # ebx points to x[0]
   f: 8b 54 24 18        mov    edx,[esp+0x18]       # edx points to y[0]
  13: 85 c9              test   ecx,ecx              # compare n to 0
  15: 74 19              je     30 <daxpy+0x30>      # if n==0, jump to Exit
  17: 31 c0              xor    eax,eax              # i = 0 (since x^x==0)
Loop:
  19: c5 fb 10 04 c3     vmovsd xmm0,[ebx+eax*8]     # xmm0 = x[i]
  1e: c4 e2 f1 a9 04 c2 vfmadd213sd xmm0,xmm1,[edx+eax*8] # xmm0 = a*x[i] + y[i]
  24: c5 fb 11 04 c2     vmovsd xmm0,xmm1,[edx+eax*8] # y[i] = a*x[i] + y[i]
  29: 83 c0 01           add    eax,0x1              # i++
  2c: 39 c1              cmp    ecx,eax              # compare i vs n
  2e: 75 e9              jne    19 <daxpy+0x19>      # if i!=n, jump to Loop
Exit:
  30: 5b                 pop    ebx                  # restore ebx
  31: c3                 ret                         # return
```

Figure 5.12: x86-32 code for DAXPY in Figure 5.7. The lack of x86-32 registers is evident in this example, with four variables allocated to memory that are in registers in the code for the other ISAs. It also demonstrates x86-32 idioms to compare a register to zero (test ecx,ecx) or to set a register to zero (xor eax,eax).

6   RV32A: Atomic Instructions

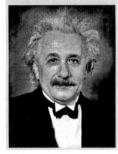

Everything should be made as simple as possible, but no simpler.

—Albert Einstein, 1933

6.1   Introduction

Our assumption is that you already understand ISA support for multiprocessing, so our job is just to explain the RV32A instructions and what they do. If you don't feel you have sufficient background or need a reminder, study "synchronization (computer science)" on Wikipedia (https://en.wikipedia.org/wiki/Synchronization_(computer_science)) or read Section 2.1 of our related RISC-V architecture book [Patterson and Hennessy 2017].

RV32A has two types of atomic operations for synchronization:

- atomic memory operations (AMO), and

- load reserved / store conditional.

Figure 6.1 is a graphical representation of the RV32A extension instruction set and Figure 6.2 lists their opcodes and instruction formats.

Figure 6.1: Diagram of the RV32A instructions.

31			25 24	20 19	15 14	12 11	7 6	0	
00010	aq	rl	00000	rs1	010	rd	0101111		R lr.w
00011	aq	rl	rs2	rs1	010	rd	0101111		R sc.w
00001	aq	rl	rs2	rs1	010	rd	0101111		R amoswap.w
00000	aq	rl	rs2	rs1	010	rd	0101111		R amoadd.w
00100	aq	rl	rs2	rs1	010	rd	0101111		R amoxor.w
01100	aq	rl	rs2	rs1	010	rd	0101111		R amoand.w
01000	aq	rl	rs2	rs1	010	rd	0101111		R amoor.w
10000	aq	rl	rs2	rs1	010	rd	0101111		R amomin.w
10100	aq	rl	rs2	rs1	010	rd	0101111		R amomax.w
11000	aq	rl	rs2	rs1	010	rd	0101111		R amominu.w
11100	aq	rl	rs2	rs1	010	rd	0101111		R amomaxu.w

Figure 6.2: RV32A opcode map has instruction layout, opcodes, format type, and names. (Table 19.2 of [Waterman and Asanović 2017] is the basis of this figure.

The AMO instructions atomically perform an operation on an operand in memory and set the destination register to the original memory value. Atomic means there can be no interrupt between the read and the write of memory, nor could other processors modify the memory value between the memory read and write of the AMO instruction.

Load reserved and store conditional provide an atomic operation across two instructions. Load reserved reads a word from memory, writes it to the destination register, and records a reservation on that word in memory. Store conditional stores a word at the address in a source register *provided there exists a load reservation on that memory address*. It writes zero to the destination register if the store succeeded, or a nonzero error code otherwise.

> **AMOs and LR/SC require naturally aligned memory addresses** because it is onerous for hardware to guarantee atomicity across cache-line boundaries.

An obvious question is: Why does RV32A have two ways to perform atomic operations? The answer is that there are two quite distinct use cases.

Programming language developers assume the underlying architecture can perform an atomic compare-and-swap operation: Compare a register value to a value in memory addressed by another register, and if they are equal, then swap a third register value with the one in memory. They make that assumption because it is a universal synchronization primitive, in that any other single-word synchronization operation can be synthesized from compare-and-swap [Herlihy 1991].

While that is powerful argument for adding such an instruction to an ISA, it requires three source registers in one instruction. Alas, going from two to three source operands would complicate the memory system interface, the integer datapath and control, and the instruction format. (The three source operands of RV32FD's multiply-add instructions affect the floating-point datapath, not the integer datapath.) Fortunately, load reserved and store conditional have only two source registers and can implement atomic compare and swap (see top half of Figure 6.3).

The rationale for also having AMO instructions is that they scale better to large multiprocessor systems than load reserved and store conditional. They can also be used to implement reduction operations efficiently. AMOs are useful as well for communicating with I/O devices, because they perform a read and a write in a single atomic bus transaction. This atomicity can both simplify device drivers and improve I/O performance. The bottom half of Figure 6.3 shows how to write a critical section using atomic swap.

Simplicity

Performance

```
# Compare-and-swap (CAS) memory word M[a0] using lr/sc.
# Expected old value in a1; desired new value in a2.
    0: 100526af    lr.w   a3,(a0)      # Load old value
    4: 06b69e63    bne    a3,a1,80     # Old value equals a1?
    8: 18c526af    sc.w   a3,a2,(a0)   # Swap in new value if so
    c: fe069ae3    bnez   a3,0         # Retry if store failed
       ... code following successful CAS goes here ...
   80:                                 # Unsuccessful CAS.


# Critical section guarded by test-and-set spinlock using an AMO.
    0: 00100293    li        t0,1         # Initialize lock value
    4: 0c55232f    amoswap.w.aq t1,t0,(a0)  # Attempt to acquire lock
    8: fe031ee3    bnez      t1,4         # Retry if unsuccessful
       ... critical section goes here ...
   20: 0a05202f    amoswap.w.rl x0,x0,(a0)  # Release lock.
```

Figure 6.3: Two examples of synchronization. The first uses load reserved/store conditional lr.w,sc.w **to implement compare-and-swap, and the second uses an atomic swap** amoswap.w **to implement a mutex.**

■ *Elaboration: Memory consistency models*

RISC-V has a relaxed memory consistency model, so other threads may view some memory accesses out of order. Figure 6.2 shows that all RV32A instructions have an *acquire bit* (aq) and a *release bit* (rl). An atomic operation with the aq bit set guarantees that other threads will see the AMO in-order with *subsequent* memory accesses. If the rl bit is set, other threads will see the atomic operation in-order with *previous* memory accesses. To learn more, [Adve and Gharachorloo 1996] is an excellent tutorial on the topic.

What's Different? The original MIPS-32 had no mechanism for synchronization, but architects added load reserved / store conditional instructions to a later MIPS ISA.

6.2   Concluding Remarks

RV32A is optional, and a RISC-V processor is simpler without it. However, as Einstein said, everything should be as simple *as possible*, but no simpler. Many situations require RV32A.

6.3   To Learn More

S. V. Adve and K. Gharachorloo. Shared memory consistency models: A tutorial. *Computer*, 29(12):66–76, 1996.

M. Herlihy. Wait-free synchronization. *ACM Transactions on Programming Languages and Systems*, 1991.

D. A. Patterson and J. L. Hennessy. *Computer Organization and Design RISC-V Edition: The Hardware Software Interface.* Morgan Kaufmann, 2017.

A. Waterman and K. Asanović, editors. *The RISC-V Instruction Set Manual, Volume I: User-Level ISA, Version 2.2.* May 2017. URL https://riscv.org/specifications/.

Notes

[1] http://parlab.eecs.berkeley.edu

7 RV32C: Compressed Instructions

E. F. Schumacher
(1911-1977) wrote this economics book that advocated human-scale, decentralized, and appropriate technologies. Translated into numerous languages, it was named one of the 100 most influential books since World War II.

Code Size

Simplicity

Small is Beautiful.

—E. F. Schumacher, 1973

7.1 Introduction

Prior ISAs significantly expanded the number of instructions and instruction formats to shrink code size: adding short instructions with two operands instead of three, small immediate fields, and so on. ARM and MIPS invented whole ISAs twice to shrink code: ARM Thumb and Thumb-2 plus MIPS16 and microMIPS. These new ISAs hampered the processor and the compiler and increased the cognitive load on the assembly language programmer.

RV32C takes a novel approach: *every* short instruction *must* map to *one* single standard 32-bit RISC-V instruction. Moreover, only the assembler and linker are aware of the 16-bit instructions, and it is up to them to replace a wide instruction with its narrow cousin. The compiler writer and assembly language programmer can be blissfully oblivious of the RV32C instructions and their formats, except ending up with programs that are smaller than most. Figure 7.1 is a graphical representation of the RV32C extension instruction set.

The RISC-V architects chose the instructions in the RVC extension to obtain good code compression across a range of programs, using three observations to fit them into 16 bits. First, ten popular registers (a0–a5, s0–s1, sp, and ra) are accessed far more than the rest. Second, many instructions overwrite one of their source operands. Third, immediate operands tend to be small, and some instructions favor certain immediates. So, many RV32C instructions can access only the popular registers; some instructions implicitly overwrite a source operand; and almost all immediates are reduced in size, with loads and stores using only unsigned offsets in multiples of the operand size.

Figures 7.3 and 7.4 list the RV32C code for Insertion Sort and DAXPY. We show the RV32C instructions to demonstrate the impact of compression explicitly, but normally these instructions are invisible in the assembly language program. The comments show the equivalent 32-bit instructions to the RV32C instructions parenthetically. Appendix A includes the 32-bit RISC-V instruction that corresponds to each 16-bit RV32C instruction.

For example, at address 4 in Insertion Sort in Figure 7.3, the assembler replaced the following 32-bit RV32I instruction:

```
addi a4,x0,1  # i = 1
```

with this 16-bit RV32C instruction:

RV32C

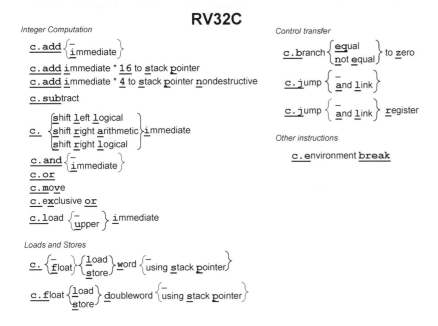

Figure 7.1: Diagram of the RV32C instructions. The immediate fields of the shift instructions and `c.addi4spn` are zero extended and sign extended for the other instructions.

```
c.li a4,1     # (expands to addi a4,x0,1) i = 1
```
The RV32C load immediate instruction is narrower because it must specify only one register and a small immediate. The `c.li` machine code is only four hexadecimal digits in Figure 7.3, showing that the `c.li` instruction is indeed 2 bytes long.

Another example is at address 10 in Figure 7.3, where the assembler replaced:
```
add a2,x0,a3  # a2 is pointer to a[j]
```
with this 16-bit RV32C instruction:
```
c.mv a2,a3    # (expands to add a2,x0,a3) a2 is pointer to a[j]
```
The RV32C move instruction is merely 16 bits long because it specifies only two registers.

While the processor designer can't ignore RV32C, an implementation trick makes them inexpensive: a decoder translates all 16-bit instructions into their equivalent 32-bit version *before* they execute. Figures 7.6 to 7.8 list the RV32C instruction formats and opcodes that the decoder translates. It is equivalent to only 400 gates when the tiniest 32-bit processor—without any RISC-V extensions—is 8000 gates. If it's 5% of such a tiny design, the decoder nearly disappears inside a moderate processor that with caches is order 100,000 gates.

What's Different? There are no byte or halfword instructions in RV32C because other instructions have a bigger influence on code size. The small size advantage of Thumb-2 over RV32C in Figure 1.5 on page 9 is due to the code size savings of Load and Store Multiple on procedure entry and exit. RV32C excludes them to maintain the one-to-one mapping to RV32G instructions, which omits them to reduce implementation complexity for high-end processors. Since Thumb-2 is a separate ISA from ARM-32, but a processor can switch between them, the hardware must have two instruction decoders: one for ARM-32 and one for Thumb-2. RV32GC is a single ISA, so RISC-V processors need only a single decoder.

Cost

Benchmark	ISA	ARM Thumb-2	microMIPS	x86-32	RV32I+RVC
Insertion Sort	Instructions	18	24	20	19
	Bytes	46	56	45	52
DAXPY	Instructions	10	12	16	11
	Bytes	28	32	50	28

Figure 7.2: Instructions and code size for Insertion Sort and DAXPY for compressed ISAs.

■ *Elaboration: Why would architects ever skip RV32C?*

Instruction decode can be a bottleneck for superscalar processors that try to fetch several instructions per clock cycle. Another example is *macrofusion*, whereby the instruction decoder
combines RISC-V instructions together to form more powerful instructions for execution (see
Chapter 1). A mix of 16-bit RV32C and 32-bit RV32I instructions can make sophisticated
decoding more difficult to complete within the clock cycle of a high-performance implementation.

7.2   Comparing RV32GC, Thumb-2, microMIPS, and x86-32

Figure 7.2 summarizes the size of Insertion Sort and DAXPY for these four ISAs.

Of the 19 original RV32I instructions in Insertion Sort, 12 become RV32C, so the code
shrinks from $19 \times 4 = 76$ bytes to $12 \times 2 + 7 \times 4 = 52$ bytes, saving $24/76 = 32\%$. DAXPY
shrinks from $11 \times 4 = 44$ bytes to $8 \times 2 + 3 \times 4 = 28$ bytes, or $16/44 = 36\%$.

The results for these two small examples are surprisingly in line with Figure 1.5 on page 9
in Chapter 2, which shows that RV32G code is about 37% larger than RV32GC code, for
a larger set of much bigger programs. To achieve that level of savings, over half of the
instructions in the programs had to be RV32C instructions.

■ *Elaboration: Is RV32C really unique?*

RV32I instructions are indistinguishable in RV32IC. Thumb-2 is actually a separate ISA
with 16-bit instructions plus most but not all of ARMv7. For example, *Compare and Branch
on Zero* is in Thumb-2 but not ARMv7, and vice versa for *Reverse Subtract with Carry*.
Nor is microMIPS32 a superset of MIPS32. For example, microMIPS multiplies branch
displacements by two but it's four in MIPS32. RISC-V *always* multiplies by two.

7.3   Concluding Remarks

I would have written a shorter letter, but I did not have the time.

—Blaise Pascal, 1656.

He was a mathematician who built one of the first mechanical calculators, which led
Turing Award laureate Niklaus Wirth to name a programming language after him.

Code Size

RV32C gives RISC-V one of the smallest code sizes today. You can almost think of
them as hardware-assisted pseudoinstructions. However, now the assembler is hiding them
from the assembly language programmer and compiler writer rather than, as in Chapter 3,

expanding the real instruction set with popular operations that make RISC-V code easier to use and to read. Both approaches aid programmer productivity.

We consider RV32C as one of RISC-V's best examples of a simple, powerful mechanism that improves its cost-performance.

Elegance

7.4 To Learn More

A. Waterman and K. Asanović, editors. *The RISC-V Instruction Set Manual, Volume I: User-Level ISA, Version 2.2*. May 2017. URL `https://riscv.org/specifications/`.

Notes

[1]`http://parlab.eecs.berkeley.edu`

```
# RV32C (19 instructions, 52 bytes)
# a1 is n, a3 points to a[0], a4 is i, a5 is j, a6 is x
  0: 00450693 addi   a3,a0,4    # a3 is pointer to a[i]
  4: 4705     c.li   a4,1       # (expands to addi a4,x0,1) i = 1
Outer Loop:
  6: 00b76363 bltu   a4,a1,c    # if i < n, jump to Continue Outer loop
  a: 8082     c.ret             # (expands to jalr x0,ra,0) return from function
Continue Outer Loop:
  c: 0006a803 lw     a6,0(a3)   # x = a[i]
 10: 8636     c.mv   a2,a3      # (expands to add a2,x0,a3) a2 is pointer to a[j]
 12: 87ba     c.mv   a5,a4      # (expands to add a5,x0,a4) j = i
InnerLoop:
 14: ffc62883 lw     a7,-4(a2)  # a7 = a[j-1]
 18: 01185763 ble    a7,a6,26   # if a[j-1] <= a[i], jump to Exit InnerLoop
 1c: 01162023 sw     a7,0(a2)   # a[j] = a[j-1]
 20: 17fd     c.addi a5,-1      # (expands to addi a5,a5,-1) j--
 22: 1671     c.addi a2,-4      # (expands to addi a2,a2,-4)decr a2 to point to a[j]
 24: fbe5     c.bnez a5,14      # (expands to bne a5,x0,14)if j!=0,jump to InnerLoop
Exit InnerLoop:
 26: 078a     c.slli a5,0x2     # (expands to slli a5,a5,0x2) multiply a5 by 4
 28: 97aa     c.add  a5,a0      # (expands to add a5,a5,a0)a5 = byte address of a[j]
 2a: 0107a023 sw     a6,0(a5)   # a[j] = x
 2e: 0705     c.addi a4,1       # (expands to addi a4,a4,1) i++
 30: 0691     c.addi a3,4       # (expands to addi a3,a3,4) incr a3 to point to a[i]
 32: bfd1     c.j    6          # (expands to jal x0,6) jump to Outer Loop
```

Figure 7.3: RV32C code for Insertion Sort. The twelve 16-bit instructions make the code 32% smaller. The width of each instruction is evident by the number of hexadecimal characters in the second column. The RV32C instructions (starting with c.) are shown explicitly in this example, but normally assembly language programmers and compilers cannot see them.

```
# RV32DC (11 instructions, 28 bytes)
# a0 is n, a1 is pointer to x[0], a2 is pointer to y[0], fa0 is a
   0: cd09      c.beqz a0,1a        # (expands to beq a0,x0,1a) if n==0, jump to Exit
   2: 050e      c.slli a0,a0,0x3    # (expands to slli a0,a0,0x3) a0 = n*8
   4: 9532      c.add a0,a2         # (expands to add a0,a0,a2) a0 = address of x[n]
Loop:
   6: 2218      c.fld fa4,0(a2)     # (expands to fld fa4,0(a2) ) fa5 = x[]
   8: 219c      c.fld fa5,0(a1)     # (expands to fld fa5,0(a1) ) fa4 = y[]
   a: 0621      c.addi a2,8         # (expands to addi a2,a2,8) a2++ (incr. ptr to y)
   c: 05a1      c.addi a1,8         # (expands to addi a1,a1,8) a1++ (incr. ptr to x)
   e: 72a7f7c3  fmadd.d fa5,fa5,fa0,fa4 # fa5 = a*x[i] + y[i]
  12: fef63c27  fsd fa5,-8(a2)      # y[i] = a*x[i] + y[i]
  16: fea618e3  bne a2,a0,6         # if i != n, jump to Loop
Exit:
  1a: 8082      ret                 # (expands to jalr x0,ra,0) return from function
```

Figure 7.4: RV32DC code for DAXPY. The eight 16-bit instructions shrink the code by 36%. The width of each instruction is evident by the number of hexadecimal characters in the second column. The RV32C instructions (starting with c.) are shown explicitly in this example, but normally they are invisible to the assembly language programmer and compiler.

15 14 13	12	11 10 9	8	7	6 5	4 3 2	1 0		
000	nzimm[5]	0			nzimm[4:0]		01	CI	c.nop
000	nzimm[5]	rs1/rd≠0			nzimm[4:0]		01	CI	c.addi
001	imm[11|4|9:8|10|6|7|3:1|5]						01	CJ	c.jal
010	imm[5]	rd≠0			imm[4:0]		01	CI	c.li
011	nzimm[9]	2			nzimm[4|6|8:7|5]		01	CI	c.addi16sp
011	nzimm[17]	rd≠{0,2}			nzimm[16:12]		01	CI	c.lui
100	nzuimm[5]	00	rs1′/rd′		nzuimm[4:0]		01	CI	c.srli
100	nzuimm[5]	01	rs1′/rd′		nzuimm[4:0]		01	CI	c.srai
100	imm[5]	10	rs1′/rd′		imm[4:0]		01	CI	c.andi
100	0	11	rs1′/rd′	00	rs2′		01	CR	c.sub
100	0	11	rs1′/rd′	01	rs2′		01	CR	c.xor
100	0	11	rs1′/rd′	10	rs2′		01	CR	c.or
100	0	11	rs1′/rd′	11	rs2′		01	CR	c.and
101	imm[11|4|9:8|10|6|7|3:1|5]						01	CJ	c.j
110	imm[8|4:3]	rs1′			imm[7:6|2:1|5]		01	CB	c.beqz
111	imm[8|4:3]	rs1′			imm[7:6|2:1|5]		01	CB	c.bnez

Figure 7.5: RV32C opcode map (bits[1 : 0] = 01) lists layout, opcodes, format, and names. rd', rs1', and rs2' refer to the 10 popular registers a0–a5, s0–s1, sp, and ra. (Table 12.5 of Waterman and Asanović 2017] is the basis of this figure.)

15 14 13	12 11 10 9 8 7 6	5	4 3 2	1 0		
000	0		0	00	CIW	*Illegal instruction*
000	nzuimm[5:4|9:6|2|3]		rd'	00	CIW	c.addi4spn
001	uimm[5:3]	rs1'	uimm[7:6]	rd'	00	CL c.fld
010	uimm[5:3]	rs1'	uimm[2|6]	rd'	00	CL c.lw
011	uimm[5:3]	rs1'	uimm[2|6]	rd'	00	CL c.flw
101	uimm[5:3]	rs1'	uimm[7:6]	rs2'	00	CL c.fsd
110	uimm[5:3]	rs1'	uimm[2|6]	rs2'	00	CL c.sw
111	uimm[5:3]	rs1'	uimm[2|6]	rs2'	00	CL c.fsw

Figure 7.6: RV32C opcode map (bits$[1:0]$ = 00) lists layout, opcodes, format, and names. rd', rs1', and rs2' refer to the 10 popular registers a0–a5, s0–s1, sp, and ra. (Table 12.4 of Waterman and Asanović 2017] is the basis of this figure.)

15 14 13	12	11 10 9 8 7	6 5 4 3 2	1 0	
000	nzuimm[5]	rs1/rd≠0	nzuimm[4:0]	10	CI c.slli
000	0	rs1/rd≠0	0	10	CI c.slli64
001	uimm[5]	rd	uimm[4:3|8:6]	10	CSS c.fldsp
010	uimm[5]	rd≠0	uimm[4:2|7:6]	10	CSS c.lwsp
011	uimm[5]	rd	uimm[4:2|7:6]	10	CSS c.flwsp
100	0	rs1≠0	0	10	CJ c.jr
100	0	rd≠0	rs2≠0	10	CR c.mv
100	1	0	0	10	CI c.ebreak
100	1	rs1≠0	0	10	CJ c.jalr
100	1	rs1/rd≠0	rs2≠0	10	CR c.add
101	uimm[5:3|8:6]		rs2	10	CSS c.fsdsp
110	uimm[5:2|7:6]		rs2	10	CSS c.swsp
111	uimm[5:2|7:6]		rs2	10	CSS c.fswsp

Figure 7.7: RV32C opcode map (bits$[1:0]$ = 10) lists layout, opcodes, format, and names. (Table 12.6 of Waterman and Asanović 2017] is the basis of this figure.)

Format	Meaning	15 14 13	12	11 10 9	8 7	6 5	4 3 2	1 0
CR	Register	funct4		rd/rs1		rs2		op
CI	Immediate	funct3	imm	rd/rs1		imm		op
CSS	Stack-relative Store	funct3		imm		rs2		op
CIW	Wide Immediate	funct3		imm			rd$'$	op
CL	Load	funct3		imm	rs1$'$	imm	rd$'$	op
CS	Store	funct3		imm	rs1$'$	imm	rs2$'$	op
CB	Branch	funct3		offset	rs1$'$	offset		op
CJ	Jump	funct3		jump target				op

Figure 7.8: Compressed 16-bit RVC instruction formats. rd', rs1', and rs2' **refer to the 10 popular registers** a0–a5, s0–s1, sp, **and** ra. **(Table 12.1 of Waterman and Asanović 2017] is the basis of this figure.)**

8 RV32V: Vector

Seymour Cray (1925-1996) was architect of the Cray-1 in 1976, the first commercially successful supercomputer using a vector architecture. The Cray-1 was a gem; it was the world's fastest computer even *without* using the vector instructions.

The Intel Multimedia Extensions (MMX) in 1997 made SIMD popular. They were embraced and expanded via Streaming SIMD Extensions (SSE) in 1999 and Advanced Vector Extensions (AVX) in 2010. MMX fame was fueled by an Intel ad campaign showing disco-dancing workers of a semiconductor line clad in technicolor clean suits (https://www.youtube.com/watch?v=paU16B-bZEA).

Isolation of Arch from Impl

Programmability

I'm all for simplicity. If it's very complicated I can't understand it.

—Seymour Cray

8.1 Introduction

This chapter focuses on *data-level parallelism*, where there is plenty of data that the desired application can compute on concurrently. Arrays are a popular example. While fundamental to scientific applications, multimedia programs use arrays as well. The former uses single- and double-precision float-point data and the latter often uses 8- and 16-bit integer data.

The best known architecture for data-level parallelism is *Single Instruction Multiple Data* (*SIMD*). SIMD first became popular by partitioning 64-bit registers into many 8-, 16-, or 32-bit pieces and then computing on them in parallel. The opcode supplied the data width and the operation. Data transfers are simply loads and stores of a single (wide) SIMD register.

The first step of partitioning existing 64-bit registers is tempting because it is straightforward. To make SIMD faster, architects subsequently widen the registers to compute more partitions concurrently. Because the SIMD ISAs belong to the incremental school of design, and the opcode specifies the data width, expanding the SIMD registers also expands the SIMD instruction set. Each subsequent step of doubling the width of SIMD registers and the number of SIMD instructions leads ISAs down the path of escalating complexity, which is borne by processor designers, compiler writers, and assembly language programmers.

An older and, in our opinion, more elegant alternative to exploit data-level parallelism is the *vector* architecture. This chapter provides our rationale for using vectors instead of SIMD in RISC-V.

Vector computers gather objects from main memory and put them into long, sequential vector registers. Pipelined execution units compute very efficiently on these vector registers. Vector architectures then scatter the results back from the vector registers to main memory. The size of the vector registers is determined by the implementation, rather than baked into the opcode, as with SIMD. As we shall see, *separating the vector length and maximum operations per clock cycle from the instruction encoding is the crux of the vector architecture:* the vector microarchitect can flexibly design the data-parallel hardware without affecting the programmer, and the programmer can take advantage of longer vectors without rewriting the code. In addition, vector architectures have many fewer instructions than SIMD architectures. Moreover, vector architectures have well-established compiler technology, unlike SIMD.

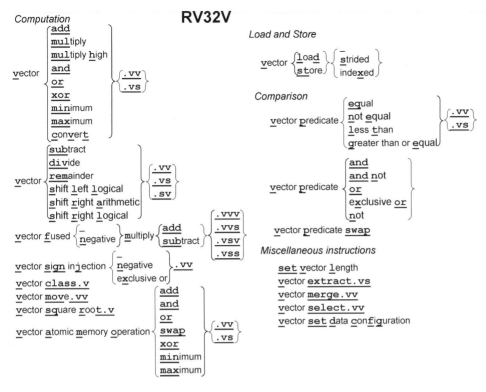

Figure 8.1: Diagram of the RV32V instructions. Because of dynamic register typing, this instruction diagram also works without change for RV64V in Chapter 9.

Vector architectures are rarer than SIMD architectures, so fewer readers know vector ISAs. Thus, this chapter will have a more tutorial flavor than earlier ones. If you want to dig deeper into vector architectures, read Chapter 4 and Appendix G of [Hennessy and Patterson 2011]. RV32V also has novel features that simplify the ISA, which requires more explanation even if you already are familiar with vector architectures.

Simplicity

8.2 Vector Computation Instructions

Figure 8.1 is a graphical representation of the RV32V extension instruction set. The RV32V encoding has not been finalized, so this edition does not include the usual instruction-layout diagram.

Virtually every integer and floating-point computation instruction from an earlier chapter has a vector version: Figure 8.1 inherits operations from RV32I, RV32M, RV32F, RV32D, and RV32A. There are several types of each vector instruction depending on whether the source operands are all vectors (.vv suffix) or a vector source operand and a scalar source operand (.vs suffix). A scalar suffix means an x or f register is an operand along with a vector register (v). For example, our DAXPY program (Figure 5.7 on page 55 in Chapter 5) calculates $Y = a \times X + Y$, where X and Y are vectors, and a is a scalar. For vector-scalar operations, the rs1 field specifies the scalar register to be accessed.

Asymmetric operations like subtraction and division offer a third variation of vector instructions where the first operand is scalar and the second is vector (.sv suffix). Operations like $Y = a - X$ use them. They are superfluous for symmetric operations like addition and multiplication, so those instructions have no .sv version. The fused multiply-add instructions have three operands, so they have the largest combination of vector and scalar options: .vvv, .vvs, .vsv, and .vss.

Readers may notice that Figure 8.1 ignores the data type and width of the vector operations. The next section explains why.

8.3   Vector Registers and Dynamic Typing

RV32V adds 32 vector registers, whose names start with v, but the number of *elements* per vector register varies. That number depends on both the width of the operations and on the amount of memory dedicated to vector registers, which is up to the processor designer. For example, if the processor allocated 4096 bytes for vector registers, that is enough for all 32 vector registers to have 16 64-bit elements, 32 32-bit elements, 64 16-bit elements, or 128 8-bit elements.

To keep the number of elements flexible in a vector ISA, a vector processor calculates the *maximum vector length* (mvl) that programs use to run properly on processors with differing amounts of memory for vector registers. The vector length register (vl) sets the number of elements in a vector for a particular operation, which helps programs when a dimension of an array is not a multiple of mvl. We'll demonstrate mvl, vl, and the eight predicate registers (vp$i$) in more detail in the following sections.

RV32V takes the novel approach of associating the data type and length with the *vector registers* rather than with the *instruction opcodes*. A program tags the vector registers with their data type and width before executing the vector computation instructions. *Dynamic register typing* slashes the number of vector instructions, important because there are often six integer and three floating-point versions of each vector instruction as Figure 8.1 shows. As we shall see in Section 8.9 when we confront the numerous SIMD instructions, a dynamically typed vector architecture reduces the cognitive load on the assembly language programmer and the difficulty of the compiler's code generator.

Programmability

Another advantage of dynamic typing is that programs can disable unused vector registers. This feature allocates all the vector memory to the enabled vector registers. For example, suppose only two vector registers are enabled, they are type 64-bit floats, and the processor has 1024 bytes of vector register memory. The processor would halve the memory, giving each vector register 512 bytes or $512/8 = 64$ elements and therefore set mvl to 64. Thus, mvl is dynamic, but its value is set by the processor and cannot be directly changed by software.

The source and destination registers determine the type and size of the operation and the result, so conversions are implicit with dynamic typing. For example, a processor can multiply a vector of double-precision floating-point numbers by a single-precision scalar without first having to convert the operands to the same precision. This bonus benefit reduces the total number of vector instructions and the number of instructions executed.

The vsetdcfg instruction sets the vector register types. Figure 8.2 shows the vector register types available to RV32V plus more types for RV64V (see Chapter 9). RV32V requires that vector floating-point operations have the scalar versions also. Thus, you must have at least RV32FV to use the F32 type and RV32FDV to use the F64 type. RV32V introduces a 16-bit floating-point format type F16. If an implementation supports both RV32V and RV32F, then it must support both F16 and F32 formats.

Type	Floating Point		Signed Integer		Unsigned Integer	
Width	Name	vetype	Name	vetype	Name	vetype
8 bits	–	–	X8	10 100	X8U	11 100
16 bits	F16	01 101	X16	10 101	X16U	11 101
32 bits	F32	01 110	X32	10 110	X32U	11 110
64 bits	F64	01 111	X64	10 111	X64U	11 111

Figure 8.2: RV32V encodings of vector register types. The rightmost three bits of the field show the width of the data, and the two leftmost bits give its type. X64 and U64 are available only for RV64V. F16 and F32 require the RV32F extension and F64 requires RV32F and RV32D. F16 is the IEEE 754-2008 16-bit floating-point format (binary16). Setting `vetype` to 00000 disables the vector registers. (Table 17.4 of [Waterman and Asanović 2017] is the basis of this figure.)

■ *Elaboration: RV32V can switch context quickly.*

One reason vector architectures were less popular than SIMD architectures was concern that adding large vector registers would stretch the time to save and restore a program on an interrupt, called a *context switch*. Dynamic register typing helps. The programmer must tell the processor which vector registers are being used, which means processor needs to save and restore only those registers on a context switch. The RV32V convention is to disable *all* vector registers when the vector instructions aren't being used, which means a processor can have the performance benefit of vector registers but pay the extra context switch time only if an interrupt occurs while the vector instructions are executing. Earlier vector architectures had to pay the worst-case context switch cost of saving and restoring all vector registers whenever an interrupt occurred.

> **Concern about slow context switch times** led Intel to avoid adding registers in the original MMX SIMD extension. It simply reused the existing floating-point registers, which meant no extra context to switch, but a program couldn't intermix floating-point and multimedia instructions.

8.4 Vector Loads and Stores

The easiest case for vector loads and stores is dealing with single-dimension arrays that are stored sequentially in memory. Vector load fills a vector register with data from sequential addresses in memory starting with the address in the `vld` instruction. The data type associated with the vector register determines the size of the data elements and the vector length register `vl` sets the number of elements to load. Vector store `vst` does the inverse operation of `vld`.

> **Each load and store has a 7-bit unsigned immediate offset** that is scaled by the element type in the destination register for loads and the source register for stores.

For example, if a0 has 1024, and the type of v0 is X32, then `vld v0, 0(a0)` will generate the addresses 1024, 1028, 1032, 1036, ... until reaching the limit set by `vl`.

For multi-dimension arrays, some accesses will not be sequential. If stored in row major order, sequential column accesses in a two-dimensional array want data elements separated by the size of the row. Vector architectures support these accesses with *strided* data transfers: `vlds` and `vsts`. While one could get the same effect as `vld` and `vst` by setting the stride to the size of the element in `vlds` and `vsts`, `vld` and `vst` guarantee that all accesses will be sequential, which makes it easier to deliver high memory bandwidth. Another reason is that providing `vld` and `vst` reduces code size and instructions executed for the common case of unit stride. These instructions specify two source registers, with one giving the starting address and the other specifying the stride in bytes.

For example, assume the starting address in a0 was address 1024, and the size of a row in a1 was 64 bytes. `vlds v0,a0,a1` would send this sequence of addresses to memory: 1024, 1088 (1024 + 1 × 64), 1152 (1024 + 2 × 64), 1216 (1024 + 3 × 64), and so on until the vector

Programmability

length register vl tells it to stop. The returning data is written into sequential elements of the destination vector register.

Programmability

Thus far, we have assumed that the program is working with dense arrays. To support sparse arrays, vector architectures offer *indexed* data transfers: vldx and vstx. One source register for these instructions refers to a vector register and the other to a scalar register. The scalar register has the starting address of the sparse array, and each element of the vector register contains the index in bytes of the nonzero elements of the sparse array.

Suppose the starting address in a0 was address 1024, and vector register v1 had these byte indices in the first 4 elements: 16, 48, 80, 160. vldx v0,a0,v1 would send this sequence of addresses to memory: 1040 (1024 + 16), 1072 (1024 + 48), 1104 (1024 + 80), 1184 (1024 + 160). It loads the returning data into sequential elements of the destination vector register.

We used sparse arrays as our motivation for indexed loads and stores, but there are many other algorithms that access data indirectly via a table of indices.

Indexed load is also called *gather* and indexed store is often named *scatter*.

8.5   Parallelism During Vector Execution

Performance

While a simple vector processor might execute one vector element at a time, element operations are independent by definition, and so a processor could theoretically compute all of them simultaneously. The widest data for RV32G is 64 bits, and today's vector processors typically execute two, four, or eight 64-bit elements per clock cycle. Hardware handles the fringe cases when the vector length is not a multiple of the number of the elements executed per clock cycle.

Like SIMD, the number of smaller data operations is the ratio of the widths of the narrow data to the wide data. Thus, a vector processor that computes 4 64-bit operations per clock cycle would normally launch 8 32-bit, 16 16-bit, and 32 8-bit operations per clock cycle.

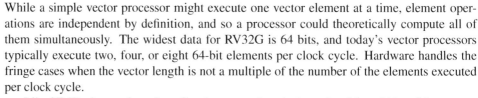
Programmability

In SIMD, the ISA architect determines the maximum number of data parallel operations per clock cycle *and* the number of elements per register. In contrast, the RV32V processor designer picks both of them without having to change the ISA or the compiler, while every doubling of SIMD register width doubles the number of SIMD instructions and requires changes to the SIMD compilers. This hidden flexibility means the identical RV32V program runs without change on the simplest and most aggressive vector processors.

8.6   Conditional Execution of Vector Operations

Performance

Some vector computations include if statements. Rather than rely on conditional branches, vector architectures include a mask that suppresses operations on some elements of a vector operation. The predicate instructions in Figure 8.1 perform conditional tests between two vectors or a vector and scalar and writes into each element of the vector mask a 1 if the condition holds or a 0 otherwise. (The vector mask must have the same number of elements as the vector registers.) Any subsequent vector instruction can use that mask, with a 1 in bit i means that element i is changed by vector operations, and a 0 means that element i is unchanged.

A program is called *vectorizable* if most operations are performed by vector instructions. Gather, scatter, and predicate instructions increase the number of vectorizable programs.

RV32V provides 8 *vector predicate registers* (vp$i$) to act as vector masks. The instructions vpand, vpandn, vpor, vpxor, and vpnot perform logical instructions to combine them together to allow efficient processing of nested conditional statements.

RV32V instructions specify either vp0 or vp1 to be the mask that controls a vector operation. To perform a normal operation on all elements, one of those two predicates registers must be set to all ones. To swap one of the other six predicate registers quickly into vp0 or vp1, RV32V has the vpswap instruction. The predicate registers are also enabled dynamically, and disabling them clears all the predicate registers quickly.

For example, suppose all the even-numbered elements of vector register v3 were negative integers and all the odd-numbered elements were positive integers. The result of this code:

```
vplt.vs    vp0,v3,x0   # set mask bits when elements of v3 < 0
add.vv,vp0 v0,v1,v2    # change elements of v0 to v1+v2 when true
```

would set all the even bits of vp0 to 1, all the odd bits to 0, and would replace all the even elements of v0 with the sum of the corresponding elements of v1 and v2. The odd elements of v0 would be unchanged.

8.7 Miscellaneous Vector Instructions

Adding to the instruction that configures the data types of vector registers mentioned above (vsetdcfg), setvl sets the vector length register (vl) and the destination register with the smaller of the source operand and the maximum vector length (mvl). The reason for picking the minimum is to decide in loops whether the vector code can run at the maximum vector length (mvl) or it must run at a smaller value to cover the remaining elements. Thus, to handle the tail, setvl is executed every loop iteration.

RV32V also has three instructions that manipulate elements within a vector register.

Vector select (vselect) produces a new result vector by gathering elements from one source data vector at the element locations specified by the second source index vector:

```
# vindices holds values from 0..mvl-1 that select elements from vsrc
vselect vdest, vsrc, vindices
```

Thus, if the first four elements of v2 contain 8, 0, 4, 2, then vselect v0,v1,v2 will replace the zeroth element of v0 with eighth element of v1, the first element of v0 with the zeroth element of v1, the second element of v0 with the fourth element of v1, and the third element of v0 with the second element of v1.

Vector merge (vmerge) resembles vector select, but it uses a vector predicate register to choose which of the sources to use. It produces a new result vector by gathering elements from one of two source registers depending on the predicate register. The new element comes from vsrc1 if the predicate vector register element is 0 or from vsrc2 if it is 1:

```
# vp0 bit i determines whether new element i for vdest
# comes from vsrc1 (if bit i == 0) or vsrc2 (if bit i == 1)
vmerge,vp0 vdest, vsrc1, vsrc2
```

Thus, if the first four elements of vp0 contain 1, 0, 0, 1, the first four elements of v1 contain 1, 2, 3, 4, and the first four elements of v2 contain 10, 20, 30, 40, then vmerge,vp0 v0,v1,v2 will make the first four elements of v0 be 10, 2, 3, 40.

The vector extract instruction takes elements starting from the middle of one vector and places these at the beginning of a second vector register:

```
# start is scalar reg holding element starting number of vsrc
vextract vdest, vsrc, start
```

```
# a0 is n, a1 is pointer to x[0], a2 is pointer to y[0], fa0 is a
 0:   li   t0, 2<<25
 4:   vsetdcfg t0            # enable 2 64b Fl.Pt. registers
loop:
 8:   setvl t0, a0           # vl = t0 = min(mvl, n)
 c:   vld   v0, a1           # load vector x
10:   slli  t1, t0, 3        # t1 = vl * 8 (in bytes)
14:   vld   v1, a2           # load vector y
18:   add   a1, a1, t1       # increment C pointer to x by vl*8
1c:   vfmadd v1, v0, fa0, v1 # v1 += v0 * fa0 (y = a * x + y)
20:   sub   a0, a0, t0       # n -= vl (t0)
24:   vst   v1, a2           # store Y
28:   add   a2, a2, t1       # increment C pointer to y by vl*8
2c:   bnez  a0, loop         # repeat if n != 0
30:   ret                    # return
```

Figure 8.3: RV32V code for DAXPY in Figure 5.7. The machine language is missing because the RV32V opcodes are yet to be defined.

For example, if vector length vl is 64 and a0 contains 32, then vextract v0,v1,a0 will copy the last 32 elements of v1 into the first 32 elements of v0.

The vextract instruction assists reductions by following a recursive-halving approach for any binary associative operator. For example, to sum all the elements of a vector register, use vector extract to copy the last half of a vector into the first half of another vector register and halve the vector length. Next, add these two vector registers together and repeat the recursive-halving with their sum until vector length equals 1. The result in the zeroth element will be the sum of all the original elements in the vector register.

Performance

8.8 Vector Example: DAXPY in RV32V

The V in RISC-V is also for vector. The RISC-V architects had extensive positive experience with vector architectures and were frustrated that SIMD dominated microprocessors. Hence, the V is for the fifth Berkeley RISC project *and* because their ISA would highlight vectors.

Figure 8.3 shows the RV32V assembly language for DAXPY (Figure 5.7 on page 55 in Chapter 5), which we'll explain a step at a time.

RV32V DAXPY starts by enabling the vector registers needed for this function. It requires only two vector registers to hold portions of x and y, which are double-precision floating-point numbers each 8 bytes wide. The first instruction creates a constant and the second writes it to the control status register that configures vector registers (vcfgd) to get two registers of type F64 (see Figure 8.2). By definition, the hardware allocates the configured registers in numerical order, yielding v0 and v1.

Let's assume our RV32V processor has 1024 bytes of memory dedicated to vector registers. The hardware allocates the memory evenly between the two vector registers, which hold double-precision floating-point numbers (8 bytes). Each vector register has $512/8 = 64$ elements, so the processor sets the maximum vector length (mvl) for this function to 64.

The first instruction in the loop sets the vector length for the following vector instructions. The instruction setvl writes the smaller of the mvl and n into vl and t0. The insight is that if the number of iterations of the loop is larger than n, the fastest the code can crunch the data is 64 values at time, so set vl to mvl. If n is smaller than mvl, then we can't read or write beyond the end of x and y, so we should compute only on the last n elements in this final iteration of the loop. `setvl` also writes to `t0` to help with later loop bookkeeping at location 10.

The instruction `vld` at address c is a vector load from the address of x in scalar register a1. It transfers `vl` elements of x from memory to `v0`. The following shift instruction `slli` multiplies the vector length by the width of the data in bytes (8) for later use in incrementing pointers to x and y.

The instruction at address 14 (`vld`) loads `vl` elements of y from memory into `v1` and the next instruction (`add`) increments the pointer to x.

The instruction at address 1c is the jackpot. `vfmadd` multiplies `vl` elements of x (`v0`) by the scalar a (`f0`) and adds each product to `vl` elements of y (`v1`) and stores those `vl` sums back into y (`v1`).

All that is to left is store the results in memory and some loop overhead. The instruction at address 20 (`sub`) decrements n (`a0`) by `vl` to record the number of operations completed in this iteration of the loop. The following instruction (`vst`) stores `vl` results into y in memory. The instruction at address 28 (`add`) increments the pointer to y and the following instruction repeats the loop if n (`a0`) is not zero. If n is zero, the final instruction `ret` returns to the calling site.

The power of vector architecture is that each iteration of this 10-instruction loop launches $3 \times 64 = 192$ memory accesses and $2 \times 64 = 128$ floating-point multiplies and additions (assuming that n is at least 64). That averages about 19 memory accesses and 13 operations per instruction. As we shall see in the next section, these ratios for SIMD are an order of magnitude worse.

Vector architectures without `setvl` have extra *strip-mining* code to set `vl` to the last n elements of the loop and to check if n is initially zero.

Performance

8.9   Comparing RV32V, MIPS-32 MSA SIMD, and x86-32 AVX SIMD

We'll now see the contrast between how SIMD and vector executes DAXPY. If you tilt your head, you can see SIMD as a restricted vector architecture with short vector registers—eight 8-bit "elements"—but it has no vector length register and no strided or indexed data transfers.

ARM-32 has a SIMD extension called NEON but it doesn't support double-precision floating-point instructions, so it doesn't help DAXPY.

MIPS SIMD. Figure 8.5 on page 83 shows the MIPS SIMD Architecture (MSA) version of DAXPY. Each MSA SIMD instruction can operate on two floating-point numbers since the MSA registers are 128 bits wide.

Unlike RV32V, because there is no vector length register, MSA requires extra bookkeeping instructions to check for problem values of n. When n is odd, there is extra code to compute a single floating-point multiply-add since MSA must operate on pairs of operands. That code is found in locations 3c to 4c in Figure 8.5. In the unlikely but possible case when n is zero, the branch at location 10 will skip the main computation loop.

If it doesn't branch around the loop, the instruction at location 18 (`splati.d`) puts copies of a in both halves of the SIMD register w2. To add scalar data in SIMD, we need to replicate it to be as wide as the SIMD register.

Inside the loop, the `ld.d` instruction at location 1c loads two elements of y into SIMD register w0 and then increments the pointer to y. It then does the a load of two elements of x into the SIMD register w1. The following instruction at location 28 increments the pointer to x. The payoff multiply-add instruction at location 2c is next.

The (delayed) branch at the end of the loop tests to see if the pointer to y has been incremented beyond the last even element of y. If it hasn't, the loop repeats. The SIMD store in the delay slot at address 34 writes the result to two elements of y.

Such bookkeeping code is considered part of `strip mining` in vector architectures. As the caption of Figure 8.5 explains, the vector length register `vl` renders such SIMD bookkeeping code moot for RV32V. Traditional vector architectures need extra code to handle the corner case of n = 0. RV32V just makes vector instructions act like nops when n = 0.

ISA	MIPS-32 MSA	x86-32 AVX2	RV32FDV
Instructions (static)	22	29	13
Bytes (static)	88	92	52
Instructions per Main Loop	7	6	10
Results per Main Loop	2	4	64
Instructions (dynamic, n=1000)	3511	1517	163

Figure 8.4: **Number of instructions and code size of DAXPY for vector ISAs. It lists number of instructions total (static), code size, number of instructions and results per loop, and number of instructions executed (n = 1000). microMIPS with MSA shrinks code size to 64 bytes and RV32FDCV reduces it to 40 bytes.**

After the main loop terminates, the code checks to see if n is odd. If so, it performs the last multiply-add using scalar instructions from Chapter 5. The final instruction returns to the calling site.

The 7-instruction loop at the heart of the MIPS MSA DAXPY code does 6 double-precision memory accesses and 4 floating-point multiplies and additions. The average is about 1 memory access and 0.5 operations per instruction.

x86 SIMD. Intel has gone through many generations of SIMD extensions, which we see in the code in Figure 8.6 on page 84. The SSE expansion to 128-bit SIMD led to the xmm registers and instructions that can use them, and the expansion to 256-bit SIMD as part of AVX created the ymm registers and their instructions.

The first group of instructions at addresses 0 to 25 load the variables from memory, make four copies of a in a 256-bit ymm registers, and tests to ensure n is at least 4 before entering the main loop. It uses two SSE and one AVX instructions. (The caption of Figure 8.6 explains how in more detail.)

The main loop does the heart of the DAXPY computation. The AVX instruction vmovapd at address 27 loads 4 elements of x into ymm0. The AVX instruction vfmadd213pd at address 2c multiplies 4 copies of a (ymm2) times 4 elements of x (ymm0), adds 4 elements of y (in memory at address ecx+edx*8), and puts the 4 sums into ymm0. The following AVX instruction at address 32, vmovapd, stores the 4 results into y. The next three instructions increment counters and repeat the loop if needed.

As was the case for MIPS MSA, the "fringe" code between addresses 3e and 57 deals with the cases when n is not a multiple of 4. It relies on three SSE instructions.

The 6 instructions of the main loop in the x86-32 AVX2 DAXPY code do 12 double-precision memory accesses and 8 floating-point multiplies and additions. They average 2 memory accesses and about 1 operation per instruction.

■ *Elaboration: The Illiac IV was the first to show the difficulty of compiling for SIMD.*

With 64 parallel 64-bit floating-point units (FPUs), the Illiac IV was planned to have more than 1 million logic gates before Moore published his law. Its architect originally predicted 1000 million floating-point operations per second (MFLOPS), but actual performance was 15 MFLOPS at best. Costs escalated from the $8M estimated in 1966 to $31M by 1972, despite the construction of only 64 of the planned 256 FPUs. The project started in 1965 but took until 1976 to run its first real application, the year the Cray-1 was unveiled. Perhaps the most infamous supercomputer, it made a top 10 list of engineering disasters [Falk 1976].

8.10 Concluding Remarks

If the code is vectorizable, the best architecture is vector.

—Jim Smith, keynote speech, International Symposium on Computer Architecture, 1994

Figure 8.4 summarizes the number of instructions and number of bytes in DAXPY of programs for RV32IFDV, MIPS-32 MSA, and x86-32 AVX2. The SIMD computation code is dwarfed by the bookkeeping code. Two-thirds to three-fourths of the code for MIPS-32 MSA and x86-32 AVX2 is SIMD overhead, either to prepare the data for the main SIMD loop or to handle the fringe elements when n is not a multiple of the number of floating-point numbers in a SIMD register.

RV32V code in Figure 8.3 doesn't need such bookkeeping code, which halves the number of instructions. Unlike SIMD, it has a vector length register, which makes the vector instructions work at any value of n. You might think RV32V would have a problem when n is 0. It doesn't because RV32V vector instructions leave everything unchanged when $vl = 0$.

Simplicity

However, the most significant difference between SIMD and vector processing is not the static code size. The SIMD instructions execute 10 to 20 times more instructions than RV32V because each SIMD loop does only 2 or 4 elements instead of 64 in the vector case. The extra instruction fetches and instruction decodes means higher energy to perform the same task.

Performance

Comparing the results in Figure 8.4 to the scalar versions of DAXPY in Figure 5.8 on page 29 in Chapter 5, we see that SIMD roughly doubles the size of the code in instructions and bytes, but the main loop is the same size. The reduction in the dynamic number of instructions executed is a factor of 2 or 4, depending on the width of the SIMD registers. However, the RV32V vector code size increases by a factor of 1.2 (with the main loop 1.4X) but the dynamic instruction count is a factor of 43 smaller!

While dynamic instruction count is a large difference, in our view that is the second most significant disparity between SIMD and vector. Lacking a vector length register explodes the number of instructions as well as the bookkeeping code. ISAs like MIPS-32 and x86-32 that follow the incrementalist doctrine must duplicate all the old SIMD instructions defined for narrower SIMD registers every time they double the SIMD width. Surely, hundreds of MIPS-32 and x86-32 instructions were created over many generations of SIMD ISAs and hundreds more are in their future. The cognitive load on the assembly language programmer of this brute force approach to ISA evolution must be overwhelming. How can one remember what vfmadd213pd means and when to use it?

In comparison, RV32V code is unaffected by the size of the memory for vector registers. Not only is RV32V unchanged if vector memory size expands, you don't even have to recompile. Since the processor supplies the value of maximum vector length mvl, the code in Figure 8.3 is untouched whether a processor raises the vector memory from 1024 bytes to, say, 4096 bytes, or drops it to 256 bytes.

Programmability

Unlike SIMD, where the ISA dictates the required hardware—and changing the ISA means changing the compiler—the RV32V ISA allows processor designers to choose the resources for data parallelism for their application without affecting the programmer or compiler. One can argue that SIMD violates the ISA design principle from Chapter 1 of isolating the architecture from implementation.

Isolation of Arch from Impl

We think the high contrast in cost-energy-performance, complexity, and ease of programming between the modular vector approach of RV32V and the incrementalist SIMD architectures of ARM-32, MIPS-32, and x86-32 might be the most persuasive argument for RISC-V.

Elegance

8.11 To Learn More

H. Falk. What went wrong V: Reaching for a gigaflop: The fate of the famed Illiac IV was shaped by both research brilliance and real-world disasters. *IEEE spectrum*, 13(10):65–70, 1976.

J. L. Hennessy and D. A. Patterson. *Computer architecture: a quantitative approach.* Elsevier, 2011.

A. Waterman and K. Asanović, editors. *The RISC-V Instruction Set Manual, Volume I: User-Level ISA, Version 2.2.* May 2017. URL https://riscv.org/specifications/.

Notes

[1] http://parlab.eecs.berkeley.edu

```
# a0 is n, a2 is pointer to x[0], a3 is pointer to y[0], $w13 is a
00000000 <daxpy>:
   0: 2405fffe  li        a1,-2
   4: 00852824  and       a1,a0,a1          # a1 = floor(n/2)*2 (mask bit 0)
   8: 000540c0  sll       t0,a1,0x3         # t0 = byte address of a1
   c: 00e81821  addu      v1,a3,t0          # v1 = &y[a1]
  10: 10e30009  beq       a3,v1,38          # if y==&y[a1] goto Fringe (t0==0 so n is 0 | 1)
  14: 00c01025  move      v0,a2             # (delay slot) v0 = &x[0]
  18: 78786899  splati.d  $w2,$w13[0]       # w2 = fill SIMD register with copies of a
Loop:
  1c: 78003823  ld.d      $w0,0(a3)         # w0 = 2 elements of y
  20: 24e70010  addiu     a3,a3,16          # increment C pointer to y by 2 Fl.Pt. numbers
  24: 78001063  ld.d      $w1,0(v0)         # w1 = 2 elements of x
  28: 24420010  addiu     v0,v0,16          # increment C pointer to x by 2 Fl.Pt. numbers
  2c: 7922081b  fmadd.d   $w0,$w1,$w2       # w0 = w0 + w1 * w2
  30: 1467fffa  bne       v1,a3,1c          # if (end of y != ptr to y) go to Loop
  34: 7bfe3827  st.d      $w0,-16(a3)       # (delay slot) store 2 elts of y
Fringe:
  38: 10a40005  beq       a1,a0,50          # if (n is even) goto Done
  3c: 00c83021  addu      a2,a2,t0          # (delay slot) a2 = &x[n-1]
  40: d4610000  ldc1      $f1,0(v1)         # f1 = y[n-1]
  44: d4c00000  ldc1      $f0,0(a2)         # f0 = x[n-1]
  48: 4c206b61  madd.d    $f13,$f1,$f13,$f0 # f13 = f1 + f0 * f13 (muladd if n is odd)
  4c: f46d0000  sdc1      $f13,0(v1)        # y[n-1] = f13 (store odd result)
Done:
  50: 03e00008  jr        ra                # return
  54: 00000000  nop                         # (delay slot)
```

Figure 8.5: MIPS-32 MSA code for DAXPY in Figure 5.7. The bookkeeping overhead of SIMD is evident when comparing this code to the RV32V code in Figure 8.3. The first part of the MIPS MSA code (addresses 0 to 18) duplicate the scalar variable a in a SIMD register and to check to ensure n is at least 2 before entering the main loop. The third part of the MIPS MSA code (addresses 38 to 4c) handle the fringe case when n is not a multiple of 2. Such bookkeeping code is unneeded in RV32V because the vector length register vl and the setvl instruction lets the loop work for all values of n, whether odd or even.

```
# eax is i, n is esi, a is xmm1, pointer to x[0] is ebx, pointer to y[0] is ecx
00000000 <daxpy>:
   0: 56                     push   esi
   1: 53                     push   ebx
   2: 8b 74 24 0c            mov    esi,[esp+0xc]    # esi = n
   6: 8b 5c 24 18            mov    ebx,[esp+0x18]   # ebx = x
   a: c5 fb 10 4c 24 10      vmovsd xmm1,[esp+0x10]  # xmm1 = a
  10: 8b 4c 24 1c            mov    ecx,[esp+0x1c]   # ecx = y
  14: c5 fb 12 d1            vmovddup xmm2,xmm1       # xmm2 = {a,a}
  18: 89 f0                  mov    eax,esi
  1a: 83 e0 fc               and    eax,0xfffffffc   # eax = floor(n/4)*4
  1d: c4 e3 6d 18 d2 01      vinsertf128 ymm2,ymm2,xmm2,0x1 # ymm2 = {a,a,a,a}
  23: 74 19                  je     3e               # if n < 4 goto Fringe
  25: 31 d2                  xor    edx,edx          # edx = 0
Loop:
  27: c5 fd 28 04 d3         vmovapd ymm0,[ebx+edx*8] # load 4 elements of x
  2c: c4 e2 ed a8 04 d1      vfmadd213pd ymm0,ymm2,[ecx+edx*8] # 4 mul adds
  32: c5 fd 29 04 d1         vmovapd [ecx+edx*8],ymm0 # store into 4 elements of y
  37: 83 c2 04               add    edx,0x4
  3a: 39 c2                  cmp    edx,eax          # compare to n
  3c: 72 e9                  jb     27               # repeat loop if < n
Fringe:
  3e: 39 c6                  cmp    esi,eax          # any fringe elements?
  40: 76 17                  jbe    59               # if (n mod 4) == 0 goto Done
FringeLoop:
  42: c5 fb 10 04 c3         vmovsd xmm0,[ebx+eax*8] # load element of x
  47: c4 e2 f1 a9 04 c1      vfmadd213sd xmm0,xmm1,[ecx+eax*8] # 1 mul add
  4d: c5 fb 11 04 c1         vmovsd [ecx+eax*8],xmm0 # store into element of y
  52: 83 c0 01               add    eax,0x1          # increment Fringe count
  55: 39 c6                  cmp    esi,eax          # compare Loop and Fringe counts
  57: 75 e9                  jne    42 <daxpy+0x42>  # repeat FringeLoop if != 0
Done:
  59: 5b                     pop    ebx              # function epilogue
  5a: 5e                     pop    esi
  5b: c3                     ret
```

Figure 8.6: x86-32 AVX2 code for DAXPY in Figure 5.7. The SSE instruction vmovsd **at address a loads a into half of the 128-bit** xmm1 **register. The SSE instruction** vmovddup **at address 14 duplicates a into both halves of** xmm1 **for later SIMD computation. The AVX instruction** vinsertf128 **at address 1d makes four copies of a in** ymm2 **starting from the two copies of a in** xmm1. **The three AVX instructions at addresses 42 to 4d** (vmovsd, vfmadd213sd, vmovsd) **handle when mod(n,4)** $\neq$ **0. They perform the DAXPY computation one element at a time, with the loop repeating until the function has performed exactly n multiple-add operations. Once again, such code is unnecessary for RV32V because the vector length register** vl **and the** setvl **instruction makes the loop work for any value of n.**

9 RV64: 64-bit Address Instructions

C. Gordon Bell (1934-) was one of the lead architects of two of the most popular minicomputer architectures of their day: the Digital Equipment Corporation PDP-11 (16-bit address), which was announced in 1970, and its successor seven years later, the Digital Equipment Corporation 32-bit address VAX-11 (Virtual Address eXtension).

There is only one mistake that can be made in computer design that is difficult to recover from—not having enough address bits for memory addressing and memory management.

—C. Gordon Bell, 1976

9.1 Introduction

Figures 9.1 to 9.4 shows graphical representations of the RV64G versions of the RV32G instructions. These figures illustrate the small increase in the number of instructions to switch to a 64-bit ISA in RISC-V. The ISAs typically add only a few word, doubleword, or long versions of the 32-bit instructions and expand all the registers, including the PC, to 64 bits. Thus, sub in RV64I subtracts two 64-bit numbers rather than two 32-bit numbers as in RV32I. RV64 is a close but actually different ISA than RV32; it adds a few instructions and the base instructions do slightly different things.

For example, Insertion Sort for RV64I in Figure 9.8 is quite near the code for RV32I in Figure 2.8 on page 27 in Chapter 2. It is the same number of instructions and the same number of bytes. The only changes are that the load and store word instructions become load and store doublewords, and the address increment goes from 4 for words (4 bytes) to 8 for doublewords (8 bytes). Figure 9.5 lists the opcodes of the RV64GC instructions in Figures 9.1 to 9.4.

Despite RV64I having 64-bit addresses and a default data size of 64 bits, 32-bit words are valid data types in programs. Hence, RV64I needs to support words just as RV32I needs to support bytes and halfwords. More specifically, since registers are now 64 bits wide, RV64I adds word versions of addition and subtraction: addw, addiw, subw. They truncate their results to 32 bits and write the sign-extended result to the destination register. RV64I also includes word versions of the shift instructions to get 32-bit shift result instead of a 64-bit shift result: sllw, slliw, srlw, srliw, sraw, sraiw. To do 64-bit data transfers, it has load and store doubleword: ld, sd. Finally, just as there are unsigned versions of load byte and load halfword in RV32I, RV64I must have an unsigned version of load word: lwu.

For similar reasons, RV64M needs to add word versions of multiply, divide, and remainder: mulw, divw, divuw, remw, remuw. To allow the programmer to synchronize on both words and doublewords, RV64A adds doubleword versions of all 11 of its instructions.

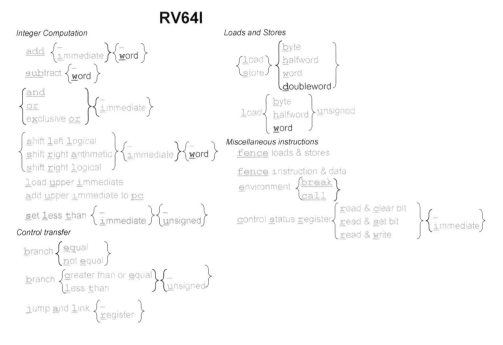

Figure 9.1: Diagram of the RV64I instructions. The underlined letters are concatenated from left to right to form RV64I instructions. The dimmed portion are the old RV64I instructions extended to 64-bit registers and the dark (red) portion are the new instructions for RV64I.

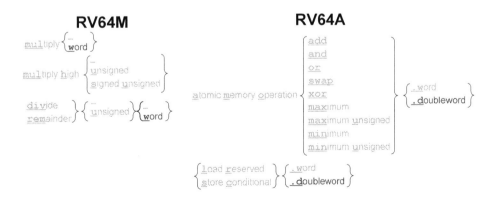

Figure 9.2: Diagrams of the RV64M and RV64A instructions.

RV64F and RV64D

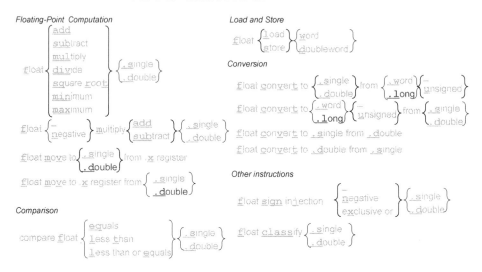

Figure 9.3: Diagram of the RV64F and RV64D instructions.

RV64C

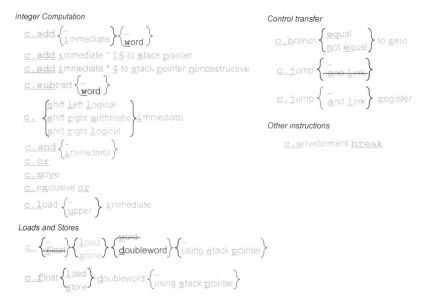

Figure 9.4: Diagram of the RV64C instructions.

31	25 24	20 19	15 14	12 11	7 6	0	
imm[11:0]		rs1	110	rd	0000011		I lwu
imm[11:0]		rs1	011	rd	0000011		I ld
imm[11:5]	rs2	rs1	011	imm[4:0]	0100011		S sd
000000	shamt	rs1	001	rd	0010011		I slli
000000	shamt	rs1	101	rd	0010011		I srli
010000	shamt	rs1	101	rd	0010011		I srai
imm[11:0]		rs1	000	rd	0011011		I addiw
0000000	shamt	rs1	001	rd	0011011		I slliw
0000000	shamt	rs1	101	rd	0011011		I srliw
0100000	shamt	rs1	101	rd	0011011		I sraiw
0000000	rs2	rs1	000	rd	0111011		R addw
0100000	rs2	rs1	000	rd	0111011		R subw
0000000	rs2	rs1	001	rd	0111011		R sllw
0000000	rs2	rs1	101	rd	0111011		R srlw
0100000	rs2	rs1	101	rd	0111011		R sraw

RV64M Standard Extension (in addition to RV32M)

0000001	rs2	rs1	000	rd	0111011	R mulw
0000001	rs2	rs1	100	rd	0111011	R divw
0000001	rs2	rs1	101	rd	0111011	R divuw
0000001	rs2	rs1	110	rd	0111011	R remw
0000001	rs2	rs1	111	rd	0111011	R remuw

RV64A Standard Extension (in addition to RV32A)

00010	aq	rl	00000	rs1	011	rd	0101111	R lr.d
00011	aq	rl	rs2	rs1	011	rd	0101111	R sc.d
00001	aq	rl	rs2	rs1	011	rd	0101111	R amoswap.d
00000	aq	rl	rs2	rs1	011	rd	0101111	R amoadd.d
00100	aq	rl	rs2	rs1	011	rd	0101111	R amoxor.d
01100	aq	rl	rs2	rs1	011	rd	0101111	R amoand.d
01000	aq	rl	rs2	rs1	011	rd	0101111	R amoor.d
10000	aq	rl	rs2	rs1	011	rd	0101111	R amomin.d
10100	aq	rl	rs2	rs1	011	rd	0101111	R amomax.d
11000	aq	rl	rs2	rs1	011	rd	0101111	R amominu.d
11100	aq	rl	rs2	rs1	011	rd	0101111	R amomaxu.d

RV64F Standard Extension (in addition to RV32F)

1100000	00010	rs1	rm	rd	1010011	R fcvt.l.s
1100000	00011	rs1	rm	rd	1010011	R fcvt.lu.s
1101000	00010	rs1	rm	rd	1010011	R fcvt.s.l
1101000	00011	rs1	rm	rd	1010011	R fcvt.s.lu

RV64D Standard Extension (in addition to RV32D)

1100001	00010	rs1	rm	rd	1010011	R fcvt.l.d
1100001	00011	rs1	rm	rd	1010011	R fcvt.lu.d
1110001	00000	rs1	000	rd	1010011	R fmv.x.d
1101001	00010	rs1	rm	rd	1010011	R fcvt.d.l
1101001	00011	rs1	rm	rd	1010011	R fcvt.d.lu
1111001	00000	rs1	000	rd	1010011	R fmv.d.x

Figure 9.5: RV64 opcode map of the base instructions and optional extensions. It shows instruction layout, opcodes, format type, and name. (Table 19.2 of [Waterman and Asanović 2017] is the basis of this figure.)

RV64F and RV64D add integer doublewords to the convert instructions, calling them *longs* so to prevent confusion with double precision floating-point data: `fcvt.l.s`, `fcvt.l.d`, `fcvt.lu.s`, `fcvt.lu.d`, `fcvt.s.l`, `fcvt.s.lu`, `fcvt.d.l`, `fcvt.d.lu`. As the integer x registers are now 64 bits wide, they can now hold double precision floating-point data, so RV64D adds two floating-point moves: `fmv.x.w` and `fmv.w.x`.

Code Size

The one exception to the superset relationship between RV64 and RV32 is the compressed instructions. RV64C replaced a few RV32C instructions, since other instructions shrank code more for 64-bit addresses. RV64C drops the compressed jump and link (`c.jal`) and the integer and floating-point load and store word instructions (`c.lw`, `c.sw`, `c.lwsp`, `c.swsp`, `c.flw`, `c.fsw`, `c.flwsp`, and `c.fswsp`). In their place, RV64C adds the more popular add and subtract word instructions (`c.addw`, `c.addiw`, `c.subw`) and load and store double-word instructions (`c.ld`, `c.sd`, `c.ldsp`, `c.sdsp`).

> ■ *Elaboration: The RV64 ABIs are lp64, lp64f, and lp64d.*
>
> lp64 means that the C language data types long and pointer are 64 bits; int is still 32 bits. The suffixes f and d indicate how floating-point arguments are passed, which is the same as for RV32 (see Chapter 3).

> ■ *Elaboration: There is no instruction diagram for RV64V*
>
> because it exactly matches RV32V due to dynamic register typing. The only change is that the X64 and X64U dynamic register types in Figure 8.2 on page 75 are available in RV64V but not RV32V.

9.2   Comparison to Other 64-bit ISAs using Insertion Sort

As Gordon Bell said at the opening of this chapter, the one fatal architecture flaw is running out of address bits. As programs pushed the limits of a 32-bit address space, architects began to make 64-bit address versions of their ISAs [Mashey 2009].

The earliest was MIPS in 1991. It extended all registers and the program counter from 32 to 64 bits and added new 64-bit versions of the MIPS-32 instructions. The MIPS-64 assembly language instructions all begin with the letter "d", such as `daddu` or `dsll` (see Figure 9.10). Programmers can mix MIPS-32 and MIPS-64 instructions in the same program. MIPS-64 dropped the load delay slot from MIPS-32 (the pipeline stalls on a read-after-write dependence).

Programmability

A decade later, it was time for a successor to x86-32. When architects increased the addressing size, they took the opportunity to make a few more improvements in x86-64:

- Increased the number of integer registers from 8 to 16 (`r8–r15`);

- Increased the number of SIMD registers from 8 to 16 (`xmm8–xmm15`); and

- Added PC-relative data addressing to better support position-independent code.

These improvements smoothed some rough edges of x86-32.

Performance

You can see the benefits by comparing the x86-32 version of Insertion Sort in Figure 2.11 on page 30 in Chapter 2 to the x86-64 version in Figure 9.11. The newer ISA keeps all the variables in registers rather than having several in memory, which reduces the instruction

ISA	ARM-64	MIPS-64	x86-64	RV64I	RV64I+RV64C
Instructions	16	24	15	19	19
Bytes	64	96	46	76	52

Figure 9.6: Number of instructions and code size for Insertion Sort for four ISAs. ARM Thumb-2 and microMIPS are 32-bit address ISAs, so are unavailable for ARM-64 and MIPS-64.

count from 20 to 15 instructions. The code size is actually larger by one byte with the newer ISA despite having fewer instructions: 46 versus 45. The reason is that to squeeze in the new opcodes to enable more registers, x86-64 added a prefix byte to identify the new instructions. The average instruction length increases in x86-64 over x86-32.

ARM faced the same address problem another decade later. Rather than evolve the old ISA to have 64-bit addresses as did x86-64, they used the opportunity to invent a brand new ISA. Given a fresh start, they changed many of the awkward ARM-32 traits to give them a modern ISA:

Programmability

- Increase the number of integer registers from 15 to 31;

- Remove the PC from the set of registers;

- Provide a register that's hardwired to zero for most instructions (r31);

- Unlike ARM-32, all ARM-64 data addressing modes work with all data sizes and types;

- ARM-64 dropped the load and store multiple instructions of ARM-32; and

- ARM-64 omitted the conditional execution option of ARM-32 instructions.

It still shares some weaknesses of ARM-32: condition codes for branch, source and destination register fields move in the instruction format, conditional move instructions, complex addressing modes, inconsistent performance counters, and only 32-bit length instructions. ARM-64 can't switch to the Thumb-2 ISA, as Thumb-2 only works with 32-bit addresses.

Unlike RISC-V, ARM decided to take a maximalist approach to ISA design. While certainly a better ISA than ARM-32, it is also bigger. For example, it has more than 1000 instructions and the ARM-64 manual is 3185 pages long [ARM 2015]. Moreover, it is still growing. There have been three expansions of ARM-64 since its announcement a few years ago.

The ARM-64 code for Insertion Sort in Figure 9.9 looks closer to the RV64I code or x86-64 code than to the ARM-32 code. For example, with 31 registers, there is no need to save and restore registers from the stack. And since the PC is no longer one of the registers, ARM-64 uses a separate return instruction.

Figure 9.6 is a table that summarizes the number of instructions and number of bytes in Insertion Sort for the ISAs. Figures 9.8 to 9.11 show the compiled code for RV64I, ARM-64, MIPS-64, and x86-64. Parenthetical phrases in the comments of these four programs identify the differences between the RV32I versions in Chapter 2 and these RV64I versions.

MIPS-64 needs the most instructions, primarily because of the nop instructions of the unfilled delayed branch slots. RV64I needs fewer because of the compare-and-branch instructions and no delayed branch. While ARM-64 and x86-64 need two compare instructions

Intel didn't invent the x86-64 ISA. When switching to 64-bit addresses, Intel invented a new ISA called Itanium that was incompatible with x86-32. Its competitor for x86-32 processors was locked out of Itanium, so AMD invented a 64-bit version of x86-32 called AMD64. Itanium eventually failed, so Intel was forced to adopt the AMD64 ISA as the 64-bit address successor of x86-32, which we call x86-64 [Kerner and Padgett 2007].

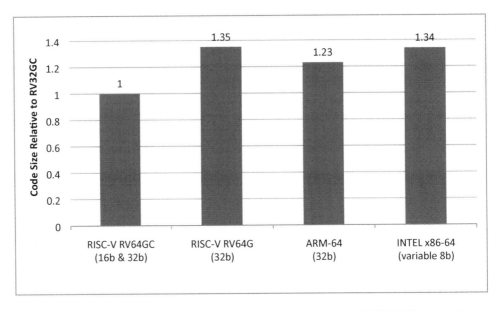

Figure 9.7: Relative program sizes for RV64G, ARM-64, and x86-64 versus RV64GC. This comparison measures much bigger programs than in Figure 9.6. This graph is the 64-bit address equivalent to the graph of 32-bit ISAs in Figure 1.5 on page 9 in Chapter 2. RV32C code size almost matches to RV64C; it is 1% smaller. There is no Thumb-2 option for ARM-64, so the core of other 64-bit ISAs significantly exceeds the size of RV64GC code. The programs measured were the SPEC CPU2006 benchmarks using the GCC compilers [Waterman 2016].

that are unnecessary for RV64I, their scaling addressing modes avoid address arithmetic instructions needed in RV64I, giving them the fewest instructions. However, RV64I+RV64C has much smaller code size, as the next section explains.

■ *Elaboration: ARM-64, MIPS-64, and x86-64 aren't the official names.*

The official names are: ARMv8 is what we call ARM-64, MIPS-IV is MIPS-64, and AMD64 is x86-64 (see the sidebar on the previous page for the history of x86-64).

9.3   Program size

Code Size

Performance

Cost

Figure 9.7 compares average relative code sizes for RV64, ARM-64, and x86-64. Compare this figure to Figure 1.5 on page 9 in Chapter 1. First, RV32GC code is almost identical in size to RV64GC; it is only 1% smaller. This closeness is also true for RV32I and RV64I. While ARM-64 code is 8% smaller than ARM-32 code, there is no 64-bit address version of Thumb-2, so all instructions remain 32-bits long. Hence, ARM-64 code is 25% *larger* than ARM Thumb-2 code. Code for x86-64 is 7% larger than x86-32 code due to adding prefix opcodes to x86-64 instructions to accommodate new operations and the expanded set of registers. RV64GC wins as ARM-64 code is 23% bigger than RV64GC and x86-64 code is 34% bigger than RV64GC. That difference is large enough that either it will improve performance due to lower instruction cache miss rates, or reduce cost by allowing a smaller instruction cache that still provides satisfactory miss rates.

9.4 Concluding Remarks

> *One of the problems of being a pioneer is you always make mistakes, and I never, never want to be a pioneer. It's always best to come second when you can look at the mistakes the pioneers made.*
>
> —Seymour Cray, architect of the first supercomputer, 1976

MIPS has its third owner. Imagination Technologies, which bought the MIPS ISA in 2012 for $100M, sold its MIPS division to Tallwood Venture Capital in 2017 for $65M.

Running out of address bits is the Achilles heel of computer architecture. Many an architecture has died from a wound there. ARM-32 and Thumb-2 remain 32-bit architectures, so they're no help for big programs. Some ISAs like MIPS-64 and x86-64 survived the transition, but x86-64 is not a paragon of ISA design and the future of MIPS-64 is unclear at the time of this writing. ARM-64 is a new large ISA, and time will tell how successful it will be.

Programmability

RISC-V benefited from designing both the 32-bit and the 64-bit architectures together, whereas older ISAs had to architect them sequentially. Unsurprisingly, the transition between 32-bit and 64-bit is easiest for RISC-V programmers and compiler writers; the RV64I ISA has virtually all RV32I instructions. Indeed, that is why we can list both RV32GCV and RV64GCV in only two pages of the Reference Card. More important, the simultaneous design meant the 64-bit architecture did not have to be squeezed into a cramped 32-bit opcode space. RV64I has plenty of room for optional instruction extensions, particularly RV64C, which makes it the leader in code size.

Room for Growth

We see the 64-bit architecture as more evidence of RISC-V's sound design, admittedly easier to achieve if you start 20 years later so that you can borrow the pioneers' good ideas as well as learn from their mistakes.

Code Size

■ *Elaboration: RV128*

RV128 began as an inside joke with the RISC-V architects, simply to show that a 128-bit address ISA was possible. However, warehouse scale computers may soon have more than 2^{64} bytes of semiconductor storage (DRAM and Flash memory), which programmers might want to access as a memory address. There are also proposals to use a 128-bit address to improve security [Woodruff et al. 2014]. The RISC-V manual does specify a full 128-bit ISA called RV128G [Waterman and Asanović 2017]. The additional instructions are basically the same as needed to go from RV32 to RV64, which Figures 9.1 to 9.4 illustrate. All the registers also grow to 128 bits, and the new RV128 instructions specify either 128-bit versions of some instructions (using Q in the name for `quadword`) or 64-bit versions of others (using D for in the name `doubleword`).

Elegance

9.5 To Learn More

I. ARM. ARMv8-A architecture reference manual. 2015.

M. Kerner and N. Padgett. A history of modern 64-bit computing. Technical report, CS Department, University of Washington, Feb 2007. URL http://courses.cs.washington.edu/courses/csep590/06au/projects/history-64-bit.pdf.

J. Mashey. The long road to 64 bits. *Communications of the ACM*, 52(1):45–53, 2009.

A. Waterman. *Design of the RISC-V Instruction Set Architecture*. PhD thesis, EECS Department, University of California, Berkeley, Jan 2016. URL http://www2.eecs.berkeley.edu/Pubs/TechRpts/2016/EECS-2016-1.html.

A. Waterman and K. Asanović, editors. *The RISC-V Instruction Set Manual, Volume I: User-Level ISA, Version 2.2.* May 2017. URL https://riscv.org/specifications/.

J. Woodruff, R. N. Watson, D. Chisnall, S. W. Moore, J. Anderson, B. Davis, B. Laurie, P. G. Neumann, R. Norton, and M. Roe. The CHERI capability model: Revisiting RISC in an age of risk. In *Computer Architecture (ISCA), 2014 ACM/IEEE 41st International Symposium on*, pages 457–468. IEEE, 2014.

Notes

[1] http://parlab.eecs.berkeley.edu

```
# RV64I (19 instructions, 76 bytes, or 52 bytes with RV64C)
# a1 is n, a3 points to a[0], a4 is i, a5 is j, a6 is x
    0: 00850693  addi  a3,a0,8    # (8 vs 4) a3 is pointer to a[i]
    4: 00100713  li    a4,1       # i = 1
Outer Loop:
    8: 00b76463  bltu  a4,a1,10   # if i < n, jump to Continue Outer loop
Exit Outer Loop:
    c: 00008067  ret              # return from function
Continue Outer Loop:
   10: 0006b803  ld    a6,0(a3)   # (ld vs lw) x = a[i]
   14: 00068613  mv    a2,a3      # a2 is pointer to a[j]
   18: 00070793  mv    a5,a4      # j = i
Inner Loop:
   1c: ff863883  ld    a7,-8(a2)  # (ld vs lw, 8 vs 4) a7 = a[j-1]
   20: 01185a63  ble   a7,a6,34   # if a[j-1] <= a[i], jump to Exit Inner Loop
   24: 01163023  sd    a7,0(a2)   # (sd vs sw) a[j] = a[j-1]
   28: fff78793  addi  a5,a5,-1   # j--
   2c: ff860613  addi  a2,a2,-8   # (8 vs 4) decrement a2 to point to a[j]
   30: fe0796e3  bnez  a5,1c      # if j != 0, jump to Inner Loop
Exit Inner Loop:
   34: 00379793  slli  a5,a5,0x3  # (8 vs 4) multiply a5 by 8
   38: 00f507b3  add   a5,a0,a5   # a5 is now byte address oi a[j]
   3c: 0107b023  sd    a6,0(a5)   # (sd vs sw) a[j] = x
   40: 00170713  addi  a4,a4,1    # i++
   44: 00868693  addi  a3,a3,8    # increment a3 to point to a[i]
   48: fc1ff06f  j     8          # jump to Outer Loop # continue outer loop
```

Figure 9.8: RV64I code for Insertion Sort in Figure 2.5. The RV64I assembly language program is very similar to the RV32I assembly language in Figure 2.8 on page 27 in Chapter 2. We list the differences in parentheses in the comments. The size of the data is now 8 bytes instead of 4, so three instructions change the constant 4 to 8. This extra width also stretches two load words (lw) to load doublewords (ld) and two store words (sw) to store doublewords (sd).

```
# ARM-64 (16 instructions, 64 bytes)
# x0 points to a[0], x1 is n, x2 is j, x3 is i, x4 is x
   0: d2800023  mov   x3, #0x1              # i = 1
Outer Loop:
   4: eb01007f  cmp   x3, x1               # compare i vs n
   8: 54000043  b.cc  10                   # if i < n, jump to Continue Outer loop
Exit Outer Loop:
   c: d65f03c0  ret                        # return from function
Continue Outer Loop:
  10: f8637804  ldr   x4, [x0, x3, lsl #3] # (x4 ca r4) vs x = a[i]
  14: aa0303e2  mov   x2, x3               # (x2 vs r2) j = i
Inner Loop:
  18: 8b020c05  add   x5, x0, x2, lsl #3   # x5 is pointer to a[j]
  1c: f85f80a5  ldur  x5, [x5, #-8]        # x5 = a[j]
  20: eb0400bf  cmp   x5, x4               # compare a[j-1] vs. x
  24: 5400008d  b.le  34                   # if a[j-1]<=a[i], jump to Exit Inner Loop

  28: f8227805  str   x5, [x0, x2, lsl #3] # a[j] = a[j-1]
  2c: f1000442  subs  x2, x2, #0x1         # j--
  30: 54ffff41  b.ne  18                   # if j != 0, jump to Inner Loop
Exit Inner Loop:
  34: f8227804  str   x4, [x0, x2, lsl #3] # a[j] = x
  38: 91000463  add   x3, x3, #0x1         # i++
  3c: 17fffff2  b     4                    # jump to Outer Loop
```

Figure 9.9: ARM-64 code for Insertion Sort in Figure 2.5. The ARM-64 assembly language program is different from to the ARM-32 assembly language in Figure 2.11 on page 30 in Chapter 2 since it is a new instruction set. The registers start with x instead of a. The data addressing modes can shift a register by 3 bits to scale the index to a byte address. With 31 registers, there is no need to save and restore registers from the stack. Since PC is not one of the registers, it uses is a separate return instruction. In fact, the code looks closer to the RV64I code or x86-64 code than to the ARM-32 code.

```
# MIPS-64 (24 instructions, 96 bytes)
# a1 is n, a3 is pointer to a[0], v0 is j, v1 is i, t0 is x
   0: 64860008 daddiu a2,a0,8   # (daddiu vs addiu, 8 vs 4) a2 is pointer to a[i]
   4: 24030001 li     v1,1      # i = 1
Outer Loop:
   8: 0065102b sltu   v0,v1,a1  # set on i < n
   c: 14400003 bnez   v0,1c     # if i < n, jump to Continue Outer Loop
  10: 00c03825 move   a3,a2     # a3 is pointer to a[j] (slot filled)
  14: 03e00008 jr     ra        # return from function
  18: 00000000 nop              # branch delay slot unfilled
Continue Outer Loop:
  1c: dcc80000 ld     a4,0(a2)  # (ld vs lw) x = a[i]
  20: 00601025 move   v0,v1     # j = i
Inner Loop:
  24: dce9fff8 ld     a5,-8(a3) # (ld vs lw, 8 vs. 4, a5 vs t1) a5 = a[j-1]
  28: 0109502a slt    a6,a4,a5  # (no load delay slot) set a[i] < a[j-1]
  2c: 11400005 beqz   a6,44     # if a[j-1] <= a[i], jump to Exit Inner Loop
  30: 00000000 nop              # branch delay slot unfilled
  34: 6442ffff daddiu v0,v0,-1  # (daddiu vs addiu) j--
  38: fce90000 sd     a5,0(a3)  # (sd vs sw, a5 vs t1) a[j] = a[j-1]
  3c: 1440fff9 bnez   v0,24     # if j != 0, jump to Inner Loop (next slot filled)
  40: 64e7fff8 daddiu a3,a3,-8  # (daddiu vs addiu, 8 vs 4) decr a2 pointer to a[j]
Exit Inner Loop:
  44: 000210f8 dsll   v0,v0,0x3 # (dsll vs sll)
  48: 0082102d daddu  v0,a0,v0  # (daddu vs addu) v0 now byte address oi a[j]
  4c: fc480000 sd     a4,0(v0)  # (sd vs sw) a[j] = x
  50: 64630001 daddiu v1,v1,1   # (daddiu vs addiu) i++
  54: 1000ffec b      8         # jump to Outer Loop (next delay slot filled)
  58: 64c60008 daddiu a2,a2,8   # (daddiu vs addiu, 8 vs 4) incr a2 pointer to a[i]
  5c: 00000000 nop              # Unncessary(?)
```

Figure 9.10: MIPS-64 code for Insertion Sort in Figure 2.5. The MIPS-64 assembly language program has several differences from to the MIPS-32 assembly language in Figure 2.10 on page 29 in Chapter 2. First, most operations for 64-bit data prepend a "d" to their names: daddiu, daddu, dsll. Like Figure 9.8, three instructions change the constant from 4 to 8 since size of the data grew from 4 to 8 bytes. Again like RV64I, the extra width also stretches two load words (lw) to load doublewords (ld) and two store words (sw) to store doublewords (sd). Finally, MIPS-64 does not have the load delay slot from MIPS-32; the pipeline stalls on a read-after-write dependence.

```
# x86-64 (15 instructions, 46 bytes)
# rax is j, rcx is x, rdx is i, rsi is n, rdi is pointer to a[0]
   0: ba 01 00 00 00 mov edx,0x1
Outer Loop:
   5: 48 39 f2          cmp rdx,rsi            # compare i vs. n
   8: 73 23             jae 2d <Exit Loop>     # if i >= n, jump to Exit Outer Loop
   a: 48 8b 0c d7       mov rcx,[rdi+rdx*8]    # x = a[i]
   e: 48 89 d0          mov rax,rdx            # j = i
Inner Loop:
  11: 4c 8b 44 c7 f8 mov r8,[rdi+rax*8-0x8]    # r8 = a[j-1]
  16: 49 39 c8          cmp r8,rcx             # compare a[j-1] vs. x
  19: 7e 09             jle 24 <Exit Loop>     # if a[j-1]<=a[i],jump to Exit InnerLoop
  1b: 4c 89 04 c7       mov [rdi+rax*8],r8     # a[j] = a[j-1]
  1f: 48 ff c8          dec rax                # j--
  22: 75 ed             jne 11 <Inner Loop>    # if j != 0, jump to Inner Loop
Exit InnerLoop:
  24: 48 89 0c c7       mov [rdi+rax*8],rcx    # a[j] = x
  28: 48 ff c2          inc rdx                # i++
  2b: eb d8             jmp 5 <Outer Loop>     # jump to Outer Loop
Exit Outer Loop:
  2d: c3                ret                    # return from function
```

Figure 9.11: x86-64 code for Insertion Sort in Figure 2.5. The x86-64 assembly language program is quite different from to the x86-32 assembly language in Figure 2.11 on page 30 in Chapter 2. First, unlike RV64I, the wider registers have different names rax, rcx, rdx, rsi, rdi, r8. Second, because x86-64 added 8 more registers, there are now enough to keep all the variables in registers instead of in memory. Third, the x86-64 instructions are longer than for x86-32 since many need to prepend 8-bits or 16-bits to fit the new instructions in the opcode space. For example, incrementing or decrementing a register (inc, dec) takes 1 byte in x86-32 but 3 bytes in x86-64. Hence, while many fewer instructions, x86-64 code size of Insertion Sort is almost identical to x86-32: 45 bytes vs. 46 bytes.

10 RV32/64 Privileged Architecture

Edsger W. Dijkstra
(1930–2002) received
the 1972 Turing Award for
fundamental contributions
to developing programming
languages.

Simplicity is prerequisite for reliability.

—Edsger W. Dijkstra

10.1 Introduction

The book so far has focused on RISC-V support for general-purpose computation: all of the instructions we've introduced are available in *user mode*, where application code usually runs. This chapter introduces two new *privilege* modes: *machine mode*, which runs the most trusted code, and *supervisor mode*, which provides support for operating systems like Linux, FreeBSD, and Windows. Both new modes are more privileged than user mode, hence the title of the chapter. More-privileged modes generally have access to all of the features of less-privileged modes, and they add additional functionality not available to less-privileged modes, such as the ability to handle interrupts and perform I/O. Processors typically spend most of their execution time in their least-privileged mode; interrupts and exceptions transfer control to more-privileged modes.

Embedded-system runtimes and operating systems use the features of these new modes to respond to external events, like the arrival of network packets; to support multitasking and protection between tasks; and to abstract and virtualize hardware features. Given the breadth of these topics, a thorough programmer's guide would be an entire additional book; instead, this chapter aims to hit the high notes of the RISC-V features. Programmers disinterested in embedded system runtimes and operating systems can either skip or skim this chapter.

Figure 10.1 is a graphical representation of the RISC-V privileged instructions, and Figure 10.2 lists these instructions' opcodes. As you can see, the privileged architecture adds

Simplicity

RV32/64 Privileged Instructions

$\left.\begin{matrix} \text{machine-mode} \\ \text{supervisor-mode} \end{matrix}\right\}$ trap `return`

supervisor-mode `fence`.virtual memory address
`wait` `for` `interrupt`

Figure 10.1: Diagram of the RISC-V privileged instructions instructions.

31	27 26 25 24	20 19	15 14	12 11	7 6	0	
0001000	00010	00000	000	00000	1110011		R sret
0011000	00010	00000	000	00000	1110011		R mret
0001000	00101	00000	000	00000	1110011		R wfi
0001001	rs2	rs1	000	00000	1110011		R sfence.vma

Figure 10.2: RISC-V privileged instruction layout, opcodes, format type, and name. (Table 6.1 of [Waterman and Asanović 2017] is the basis of this figure.)

very few instructions; instead, several new control and status registers (CSRs) expose the additional functionality.

This chapter describes the RV32 and RV64 privileged architectures together. Some concepts differ only in the size of an integer register, so to keep the descriptions concise, we introduce the term XLEN to refer to the width of an integer register in bits. XLEN is 32 for RV32 or 64 for RV64.

10.2 Machine Mode for Simple Embedded Systems

Machine mode, abbreviated as M-mode, is the most privileged mode that a RISC-V *hart* (hardware thread) can execute in. Harts running in M-mode have full access to memory, I/O, and low-level system features necessary to boot and configure the system. As such, it is the only privilege mode that all standard RISC-V processors implement; indeed, simple RISC-V microcontrollers support *only* M-mode. Such systems are the focus of this section.

The most important feature of machine mode is the ability to intercept and handle *exceptions*: unusual runtime events. RISC-V classifies exceptions into two categories. *Synchronous exceptions* arise as a result of instruction execution, as when accessing an invalid memory address or executing an instruction with an invalid opcode. *Interrupts* are external events that are asynchronous with the instruction stream, like a mouse button click. Exceptions in RISC-V are *precise*: all instructions prior to the exception completely execute, and none of the subsequent instructions appear to have begun execution. Figure 10.3 lists the standard exception causes.

Five kinds of synchronous exceptions can happen during M-mode execution:

- *Access fault exceptions* arise when a physical memory address doesn't support the access type—for example, attempting to store to a ROM.

- *Breakpoint exceptions* arise from executing an `ebreak` instruction, or when an address or datum matches a debug trigger.

- *Environment call exceptions* arise from executing an `ecall` instruction.

- *Illegal instruction exceptions* result from decoding an invalid opcode.

- *Misaligned address exceptions* occur when the effective address isn't divisible by the access size—for example, `amoadd.w` with an address of 0x12.

If you recall Chapter 2's claim that misaligned loads and stores are permitted, you might be wondering why misaligned load and store address exceptions are listed in Figure 10.3. There are two reasons. First, the atomic memory operations in Chapter 6 require naturally aligned addresses. Second, some implementors choose to omit hardware support for

Hart is a contraction of hardware thread. We use the term to distinguish them from software threads, which most programmers are familiar with. Software threads are time-multiplexed on harts. Most processor cores have only one hart.

Isolation of Arch from Impl

Misaligned instruction address exceptions can't occur with the C extension because it's never possible to jump to an odd address: branches and JAL immediates are always even, and JALR masks off the least-significant bit of its effective address. Without the C extension, this exception occurs when jumping to an address that equals 2 mod 4.

Interrupt / Exception mcause[XLEN-1]	Exception Code mcause[XLEN-2:0]	Description
1	1	Supervisor software interrupt
1	3	Machine software interrupt
1	5	Supervisor timer interrupt
1	7	Machine timer interrupt
1	9	Supervisor external interrupt
1	11	Machine external interrupt
0	0	Instruction address misaligned
0	1	Instruction access fault
0	2	Illegal instruction
0	3	Breakpoint
0	4	Load address misaligned
0	5	Load access fault
0	6	Store address misaligned
0	7	Store access fault
0	8	Environment call from U-mode
0	9	Environment call from S-mode
0	11	Environment call from M-mode
0	12	Instruction page fault
0	13	Load page fault
0	15	Store page fault

Figure 10.3: RISC-V exception and interrupt causes. The most-significant bit of mcause is set to 1 for interrupts or 0 for synchronous exceptions, and the least-significant bits identify the interrupt or exception. Supervisor interrupts and page-fault exceptions are only possible when supervisor mode is implemented (see Section 10.5). (Table 3.6 of [Waterman and Asanović 2017] is the basis of this figure.)

XLEN-1	XLEN-2		23	22	21	20	19	18	17
SD	*Reserved*		TSR	TW	TVM	MXR	SUM	MPRV	
1	XLEN-24		1	1	1	1	1	1	

16 15	14 13	12 11	10 9	8	7	6	5	4	3	2	1	0
XS	FS	MPP	*Res.*	SPP	MPIE	*Res.*	SPIE	*Res.*	MIE	*Res.*	SIE	*Res.*
2	2	2	2	1	1	1	1	1	1	1	1	1

Figure 10.4: The `mstatus` CSR. The only fields present in simple processors with only Machine mode and without the F and V extensions are the global interrupt enable, MIE, and MPIE, which after an exception holds the old value of MIE. XLEN is 32 for RV32, or 64 for RV64. Figure 3.7 of [Waterman and Asanović 2017] is the basis of this figure; see Section 3.1 of that document for a description of the other fields.

XLEN-1 12	11	10	9	8	7	6	5	4	3	2	1	0
Reserved	MEIP	*Res.*	SEIP	*Res.*	MTIP	*Res.*	STIP	*Res.*	MSIP	*Res.*	SSIP	*Res.*
Reserved	MEIE	*Res.*	SEIE	*Res.*	MTIE	*Res.*	STIE	*Res.*	MSIE	*Res.*	SSIE	*Res.*
XLEN-12	1	1	1	1	1	1	1	1	1	1	1	1

Figure 10.5: Machine interrupt CSRs. They are XLEN-bit read/write registers that hold pending interrupts (`mip`) and the interrupt enable bits (`mie`). Only the bits corresponding to lower-privilege software interrupts (SSIP), timer interrupts (STIP), and external interrupts (SEIP) in `mip` are writable through this CSR address; the remaining bits are read-only.

misaligned regular loads and stores, because it is a difficult feature to implement and is infrequently used. Processors without this hardware rely instead upon an exception handler to trap and emulate misaligned loads and stores in software, using a sequence of smaller, aligned loads and stores. Application code is none the wiser: misaligned memory accesses work as expected, albeit slowly, while the hardware remains simple. Alternatively, more performant processors can implement misaligned loads and stores in hardware. This implementation flexibility owes to RISC-V's decision to permit misaligned loads and stores using the regular load and store opcodes, following Chapter 1's guideline to isolate architecture from implementation.

Isolation of Arch from Impl

There are three standard sources of interrupts: software, timer, and external. Software interrupts are triggered by storing to a memory-mapped register and are generally used by one hart to interrupt another hart, a mechanism other architectures refer to as an *interprocessor interrupt*. Timer interrupts are raised when a hart's time comparator, a memory-mapped register named `mtimecmp`, matches or exceeds the real-time counter `mtime`. External interrupts are raised by a platform-level interrupt controller, to which most external devices are attached. As different hardware platforms have different memory maps and demand divergent features of their interrupt controllers, the mechanisms for raising and clearing these interrupts differ from platform to platform. What *is* constant across all RISC-V systems is how exceptions are handled and interrupts are masked, the topic of the next section.

10.3 Machine-Mode Exception Handling

Eight control and status registers (CSRs) are integral to machine-mode exception handling:

XLEN-1	XLEN-2		0
Interrupt		Exception Code	
1		XLEN-1	

Figure 10.6: Machine and supervisor cause CSRs (`mcause` and `scause`). When a trap is taken, the CSR is written with a code indicating the event that caused the trap. The Interrupt bit is set if the trap was caused by an interrupt. The Exception Code field contains a code identifying the last exception. Figure 10.3 maps the code values to the reason for the traps.

XLEN-1	2 1	0
BASE[XLEN-1:2]	MODE	
XLEN-2	2	

Figure 10.7: Machine and supervisor trap-vector base-address CSRs (`mtvec` and `stvec`). They are XLEN-bit read/write registers that hold trap vector configuration, consisting of a vector base address (BASE) and a vector mode (MODE). The value in the BASE field must always be aligned on a 4-byte boundary. MODE = 0 means all exceptions set the PC to BASE. MODE = 1 sets the PC to $(BASE + (4 \times cause))$ on asynchronous interrupts.

XLEN-1	0
Trap Value register [m/s]`tval`	
Exception PC register [m/s]`epc`	
Scratch register for Trap Handlers [m/s]`scratch`	
XLEN	

Figure 10.8: CSRs associated with exceptions and interrupts. Trap Value registers (`mtval` and `stval`) hold useful additional trap information such as the faulting address or an illegal instruction. The Exception PCs (`mepc` and `sepc`) point to the faulting instruction. The scratch registers (`mscratch` and `sscratch`) give trap handlers one free register to use.

Encoding	Name	Abbreviation
00	User	U
01	Supervisor	S
11	Machine	M

Figure 10.9: RISC-V privilege levels and their encoding.

- mstatus, *Machine Status*, holds the global interrupt enable, along with a plethora of other state, as Figure 10.4 shows.

- mip, *Machine Interrupt Pending*, lists the interrupts currently pending (Figure 10.5).

- mie, *Machine Interrupt Enable*, lists which interrupts the processor can take and which it must ignore (Figure 10.5).

- mcause, *Machine Exception Cause*, indicates which exception occurred (Figure 10.6).

- mtvec, *Machine Trap Vector*, holds the address the processor jumps to when an exception occurs (Figure 10.7).

- mtval, *Machine Trap Value*, holds additional trap information: the faulting address for address exceptions, the instruction itself for illegal instruction exceptions, and zero for other exceptions (Figure 10.8).

- mepc, *Machine Exception PC*, points to the instruction where the exception occurred (Figure 10.8).

- mscratch, *Machine Scratch*, holds one word of data for temporary storage for trap handlers (Figure 10.8).

When executing in M-mode, interrupts are only taken if the global interrupt-enable bit, mstatus.MIE, is set. Furthermore, each interrupt has its own enable bit in the mie CSR. The bit positions in mie correspond to the interrupt codes in Figure 10.3: for example, mie[7] corresponds to the M-mode timer interrupt. The mip CSR has the same layout and indicates which interrupts are currently pending. Putting all three CSRs together, a machine timer interrupt can be taken if mstatus.MIE=1, mie[7]=1, and mip[7]=1.

When a hart takes an exception, the hardware atomically undergoes several state transitions:

- The PC of the exceptional instruction is preserved in mepc, and the PC is set to mtvec. (For synchronous exceptions, mepc points to the instruction that caused the exception; for interrupts, it points where execution should resume after the interrupt is handled.)

- mcause is set to the exception cause, as encoded in Figure 10.3, and mtval is set to the faulting address or some other exception-specific word of information.

- Interrupts are disabled by setting MIE=0 in the mstatus CSR, and the previous value of MIE is preserved in MPIE.

- The pre-exception privilege mode is preserved in mstatus' MPP field, and the privilege mode is changed to M. Figure 10.9 shows the encoding of the MPP field. (If the processor only implements M-mode, this step is effectively skipped.)

RISC-V also supports vectored interrupts, wherein the processor jumps to an interrupt-specific address, rather than a single entry point. This addressing eliminates the need to read and decode mcause, speeding up interrupt handling. Setting mtval[0] to 1 enables this feature; interrupt cause *x* then sets the PC to (mtval-1+4*x*), instead of the usual mtval.

To avoid overwriting the contents of the integer registers, the prologue of an interrupt handler usually begins by swapping an integer register (say, a0) with the mscratch CSR. Usually, the software will have arranged for mscratch to contain a pointer to additional in-memory scratch space, which the handler uses to save as many integer registers as its body will use. After the body executes, the epilogue of an interrupt handler restores the registers it saved to memory, then again swaps a0 with mscratch, restoring both registers to their pre-exception values. Finally, the handler returns with mret, an instruction unique to M-mode. mret sets the PC to mepc, restores the previous interrupt-enable setting by copying the mstatus MPIE field to MIE, and sets the privilege mode to the value in mstatus' MPP field, essentially reversing the actions described in the preceding paragraph.

Figure 10.10 shows RISC-V assembly code for a basic timer interrupt handler following this pattern. It simply increments the time comparator then returns to the previous task, whereas a more realistic timer interrupt handler might invoke a scheduler to switch between tasks. It is not preemptible, so it keeps interrupts disabled throughout the handler. Those caveats aside, it is a complete example of a RISC-V interrupt handler on a single page!

Simplicity

Sometimes it's desirable to take a higher-priority interrupt while processing a lower-priority exception. Alas, there's only one copy of the mepc, mcause, mtval, and mstatus CSRs; taking a second interrupt would destroy the old values in these registers, causing data loss without some additional help from software. A preemptible interrupt handler can save these registers to an in-memory stack before enabling interrupts, then, just prior to exiting, disable interrupts and restore the registers from the stack.

Programmability

In addition to the mret instruction we introduced above, M-mode provides just one other instruction: wfi (*Wait For Interrupt*). wfi informs the processor that there isn't any useful work to do, so it should enter a lower-power mode until any enabled interrupt becomes pending, i.e., (mie & mip)≠0. RISC-V processors implement this instruction in a variety of ways, including stopping the clock until an interrupt becomes pending; some simply execute it as a nop. Hence, wfi is typically used inside a loop.

■ *Elaboration: wfi works whether or not interrupts are globally enabled.*

If wfi is executed when interrupts are globally enabled (mstatus.MIE=1), and then an enabled interrupt becomes pending, the processor jumps to the exception handler. If, on the other hand, wfi is executed when interrupts are globally disabled, and then an enabled interrupt becomes pending, the processor continues executing the code following the wfi. This code typically examines the mip CSR to decide what to do next. This strategy can reduce interrupt latency as compared to jumping to the exception handler, because there's no need to save and restore integer registers.

10.4   User Mode and Process Isolation in Embedded Systems

Although Machine mode is sufficient for simple embedded systems, it is only suitable when the entire codebase is trusted, since M-mode has unfettered access to the hardware platform. More often, it is not practical to trust all of the application code, because it is not known in advance or is too vast to prove correct. So, RISC-V provides mechanisms to protect the system from the untrusted code, and to protect untrusted processes from each other.

Programmability

Untrusted code must be forbidden from executing privileged instructions, like mret, and accessing privileged CSRs, like mstatus, as these would allow the program to take control of the system. This restriction is accomplished easily enough: an additional privilege mode,

```
# save registers
csrrw a0, mscratch, a0   # save a0; set a0 = &temp storage
sw a1, 0(a0)             # save a1
sw a2, 4(a0)             # save a2
sw a3, 8(a0)             # save a3
sw a4, 12(a0)            # save a4

# decode interrupt cause
csrr a1, mcause          # read exception cause
bgez a1, exception       # branch if not an interrupt
andi a1, a1, 0x3f        # isolate interrupt cause
li a2, 7                 # a2 = timer interrupt cause
bne a1, a2, otherInt     # branch if not a timer interrupt

# handle timer interrupt by incrementing time comparator
la a1, mtimecmp          # a1 = &time comparator
lw a2, 0(a1)             # load lower 32 bits of comparator
lw a3, 4(a1)             # load upper 32 bits of comparator
addi a4, a2, 1000        # increment lower bits by 1000 cycles
sltu a2, a4, a2          # generate carry-out
add a3, a3, a2           # increment upper bits
sw a3, 4(a1)             # store upper 32 bits
sw a4, 0(a1)             # store lower 32 bits

# restore registers and return
lw a4, 12(a0)            # restore a4
lw a3, 4(a0)             # restore a3
lw a2, 4(a0)             # restore a2
lw a1, 0(a0)             # restore a1
csrrw a0, mscratch, a0   # restore a0; mscratch = &temp storage
mret                     # return from handler
```

Figure 10.10: RISC-V code for a simple timer interrupt handler. The code assumes that interrupts are globally enabled by setting mstatus.MIE; that timer interrupts have been enabled by setting mie[7]; that the mtvec CSR has been set to the address of this handler; and that the mscratch CSR has been set to the address of a buffer that contains 16 bytes of temporary storage to save the registers. The prologue saves five registers, preserving a0 in mscratch and a1–a4 in memory. It then decodes the exception cause by examining mcause: interrupt if mcause<0, or synchronous exception if mcause≥0. If it is an interrupt, it checks that the lower bits of mcause equal 7, indicating an M-mode timer interrupt. If it is a timer interrupt, it adds 1000 cycles to the time comparator, so that the next timer interrupt will occur about 1000 timer cycles in the future. Finally, the epilogue restores the a0–a4 and mscratch, then returns whence it came using mret.

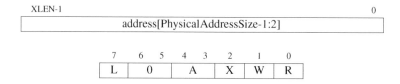

Figure 10.11: A PMP address and configuration register. The address register is right-shifted by 2, and if physical addresses are less than XLEN-2 bits wide, the upper bits are zeros. The R, W, and X fields grant read, write, and execute permissions. The A field sets the PMP mode, and the L field locks the PMP and corresponding address registers.

User mode (U-mode), denies access to these features, generating an illegal instruction exception when attempting to use an M-mode instruction or CSR. Otherwise, U-mode and M-mode behave very similarly. M-mode software can enter U-mode by setting `mstatus.MPP` to U (which, as Figure 10.9 shows, is encoded as 0), then executing an `mret` instruction. If an exception occurs in U-mode, control is returned to M-mode.

Untrusted code must also be restricted to access only its own memory. Processors that implement M and U modes have a feature called *Physical Memory Protection* (PMP), which allows M-mode to specify which memory addresses U-mode can access. PMP consists of several address registers (usually eight to sixteen) and corresponding configuration registers, which grant or deny read, write, and execute permissions. When a processor in U-mode attempts to fetch an instruction, or execute a load or store, the address is compared against all of the PMP address registers. If the address is greater than or equal to PMP address i, but less than PMP address $i+1$, then PMP $i+1$'s configuration register decides whether that access may proceed; otherwise, it raises an access exception.

Figure 10.11 shows the layout of a PMP address and configuration register. Both are CSRs, with the address registers named `pmpaddr0` to `pmpaddrN`, where $N+1$ is the number of PMPs implemented. The address registers are shifted right two bits because PMPs have a four-byte granularity. The configuration registers are densely packed in the CSRs to accelerate context switching, as Figure 10.12 shows. A PMP's configuration consists of R, W, and X bits, which when set permit loads, stores, and fetches, respectively, and a mode field, A, which when 0 disables this PMP or when 1 enables it. The PMP configuration also supports other modes and can be locked, features described in [Waterman and Asanović 2017].

10.5 Supervisor Mode for Modern Operating Systems

The PMP scheme described in the previous section is attractive for embedded systems because it provides memory protection at relatively low cost, but it has several drawbacks that limit its use in general-purpose computing. Since PMP supports only a fixed number of memory regions, it doesn't scale to complex applications. And since these regions must be contiguous in physical memory, the system can suffer from memory fragmentation. Finally, PMP doesn't efficiently support paging to secondary storage.

Fragmentation occurs when memory is available, but not in large enough contiguous chunks to be useful.

More sophisticated RISC-V processors handle these problems the same way as nearly all general-purpose architectures: using page-based virtual memory. This feature forms the core of *supervisor mode* (S-mode), an optional privilege mode designed to support modern Unix-like operating systems, such as Linux, FreeBSD, and Windows. S-mode is more privileged

31–24	23–16	15–8	7–0	
PMP3	PMP2	PMP1	PMP0	`pmpcfg0`
PMP7	PMP6	PMP5	PMP4	`pmpcfg1`
PMP11	PMP10	PMP9	PMP8	`pmpcfg2`
PMP15	PMP14	PMP13	PMP12	`pmpcfg3`

RV32

63–56	55–48	47–40	39–32	31–24	23–16	15–8	7–0	
PMP7	PMP6	PMP5	PMP4	PMP3	PMP2	PMP1	PMP0	`pmpcfg0`
PMP15	PMP14	PMP13	PMP12	PMP11	PMP10	PMP9	PMP8	`pmpcfg2`

RV64

Figure 10.12: The layout of PMP configurations in the `pmpcfg` CSRs. For RV32 (above), the sixteen configuration registers are packed into four CSRs. For RV64 (below), they are packed into the two even-numbered CSRs.

XLEN-1 … 0
Exception Delegation register `[m/s]edeleg`
Interrupt Delegation register `[m/s]ideleg`
XLEN

Figure 10.13: The delegation CSRs. Machine and supervisor exception and interrupt delegation CSRs (`medeleg`, `sedeleg`, `mideleg`, `sideleg`). They enable delegation to a lower-privilege trap handler, with the index of the bit position enabling the corresponding exception or interrupt in the [m/s]ip register.

than U-mode, but less-privileged than M-mode. Like U-mode, S-mode software can't use M-mode CSRs and instructions, and is subject to PMP restrictions. This section covers S-mode interrupts and exceptions, and the next section details the S-mode virtual-memory system.

By default, all exceptions, regardless of privilege mode, transfer control to the M-mode exception handler. Most exceptions in a Unix system, though, should invoke the operating system, which runs in S-mode. The M-mode exception handler could re-route exceptions to S-mode, but this extra code would slow down the handling of most exceptions. So, RISC-V provides an *exception delegation* mechanism, by which interrupts and synchronous exceptions can be delegated to S-mode selectively, bypassing M-mode software altogether.

The `mideleg` (*Machine Interrupt Delegation*) CSR controls which interrupts are delegated to S-mode (Figure 10.13). Like `mip` and `mie`, each bit in `mideleg` corresponds to the exception code of the same number in Figure 10.3. For example, `mideleg`[5] corresponds to the S-mode timer interrupt; if set, S-mode timer interrupts will transfer control to the S-mode exception handler, rather than the M-mode exception handler.

Any interrupt delegated to S-mode can be masked by S-mode software. The `sie` (*Supervisor Interrupt Enable*) and `sip` (*Supervisor Interrupt Pending*) CSRs are S-mode CSRs that are subsets of the `mie` and `mip` CSRs (Figure 10.14). They have the same layout as

Why not unconditionally delegate interrupts to S-mode? One reason is virtualization: if M-mode wants to virtualize a device for S-mode, its interrupts should go to M-mode, not S-mode.

S-mode doesn't directly control timer and software interrupts but instead uses the `ecall` instruction to request M-mode to set up timers or send interprocessor interrupts on its behalf. This software convention is part of the *Supervisor Binary Interface*.

XLEN-1 ... 10	9	8	7 6	5	4	3 2	1	0
Reserved	SEIP	*Res.*	*Res.*	STIP	*Res.*	*Res.*	SSIP	*Res.*
Reserved	SEIE	*Res.*	*Res.*	STIE	*Res.*	*Res.*	SSIE	*Res.*
XLEN-10	1	1	2	1	1	2	1	1

Figure 10.14: Supervisor interrupt CSRs. They are XLEN-bit read/write registers that hold pending interrupts (sip) and the interrupt enable bits (sie).

XLEN-1	XLEN-2 ...	20	19	18	17
SD	*Reserved*		MXR	SUM	*Res.*
1	XLEN-21		1	1	1

16 15	14 13	12 9	8	7 6	5	4	3 2	1	0
XS[1:0]	FS[1:0]	*Res.*	SPP	*Res.*	SPIE	UPIE	*Res.*	SIE	UIE
2	2	4	1	2	1	1	2	1	1

Figure 10.15: The sstatus CSR. sstatus is a subset of mstatus (Figure 10.4), hence the similar layout. SIE and SPIE hold the current and pre-exception interrupt enables, analogous to MIE and MPIE in mstatus. XLEN is 32 for RV32, or 64 for RV64. Figure 4.2 of [Waterman and Asanović 2017] is the basis of this figure; see Section 4.1 of that document for a description of the other fields.

their M-mode counterparts, but only the bits corresponding to interrupts that have been delegated in mideleg are readable and writable through sie and sip. The bits corresponding to interrupts that haven't been delegated are always zero.

M-mode can also delegate synchronous exceptions to S-mode using the medeleg (*Machine Exception Delegation*) CSR (Figure 10.13). The mechanism is analogous to interrupt delegation, but the bits in medeleg correspond instead to the synchronous exception codes in Figure 10.3. For example, setting medeleg[15] will delegate store page faults to S-mode.

Note that exceptions will never transfer control to a less-privileged mode, no matter the delegation settings. An exception that occurs in M-mode is always handled in M-mode. An exception that occurs in S-mode might be handled by either M-mode or S-mode, depending on the delegation configuration, but never U-mode.

S-mode has several exception-handling CSRs, scause, stvec, sepc, stval, sscratch, and sstatus, which perform the same function as their M-mode counterparts described in Section 10.2 (Figures 10.7 to 10.8). Figure 10.15 shows the layout of the sstatus register. The supervisor exception return instruction, sret, behaves the same as mret, but it acts on the S-mode exception-handling CSRs instead of the M-mode ones.

The act of taking an exception is also very similar to M-mode. If a hart takes an exception and it is delegated to S-mode, the hardware atomically undergoes several similar state transitions, using S-mode CSRs instead of M-mode ones:

Simplicity

- The PC of the exceptional instruction is preserved in sepc, and the PC is set to stvec.

- scause is set to the exception cause, as encoded in Figure 10.3, and stval is set to the faulting address or some other exception-specific word of information.

- Interrupts are disabled by setting SIE=0 in the sstatus CSR, and the previous value

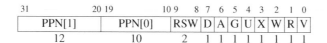

Figure 10.16: An RV32 Sv32 page-table entry (PTE).

of SIE is preserved in SPIE.

- The pre-exception privilege mode is preserved in `sstatus`' SPP field, and the privilege mode is changed to S.

10.6 Page-Based Virtual Memory

S-mode provides a conventional virtual memory system that divides memory into fixed-size *pages* for the purposes of address translation and memory protection. When paging is enabled, most addresses (including load and store effective addresses and the PC) are *virtual addresses* that must be translated into *physical addresses* in order to access physical memory. Virtual addresses are translated to physical addresses by traversing a high-radix tree known as the *page table*. A leaf node in the page table indicates whether the virtual address maps to a physical page, and, if so, which privilege modes and access types have permission to access the page. Accessing a page that is unmapped or grants insufficient permissions results in a *page fault exception*.

RISC-V paging schemes are named Sv*X*, where *X* is the size of a virtual address in bits. RV32's paging scheme, Sv32, supports a 4 GiB virtual-address space, which is divided into 2^{10} *megapages* of size 4 MiB. Each megapage is subdivided into 2^{10} *base pages*—the fundamental unit of paging—each 4 KiB. Hence, Sv32's page table is a two-level tree of radix 2^{10}. Each entry in the page table is four bytes, so a page table is itself 4 KiB. It's no coincidence that a page table is exactly the size of a page: this design simplifies operating-system memory allocation.

Figure 10.16 shows the layout of an Sv32 page-table entry (PTE), which has the following fields, explained from right to left:

- The V bit indicates whether the rest of this PTE is valid (V=1). If V=0, any virtual-address translation that traverses this PTE results in a page fault.

- The R, W, and X bits indicate whether the page has read, write, and execute permissions, respectively. If all three bits are 0, this PTE is a pointer to the next level of the page table; otherwise, it's a leaf of the tree.

- The U bit indicates whether this page is a user page. If U=0, U-mode cannot access this page, but S-mode can. If U=1, U-mode can access this page, but S-mode cannot.

- The G bit indicates this mapping exists in all virtual-address spaces, information the hardware can use to improve address-translation performance. It is typically only used for pages that belong to the operating system.

- The A bit indicates whether the page has been accessed since the last time the A bit was cleared.

> **4 KiB pages have been popular for five decades** starting with the IBM 360 model 67. Atlas, the first computer with paging, had 3 KiB pages (it had 6-byte words). We find it remarkable that, after a half-century of exponential growth in computer performance and memory capacity, the page size remains virtually unchanged.

> **The OS relies on the A and D bits to decide which pages to swap to secondary storage.** Periodically clearing the A bits helps the OS approximate which pages have been least-recently used. The D bit indicates a page is even more expensive to swap out, because it must be written back to secondary storage.

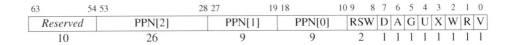

Figure 10.17: An RV64 Sv39 page-table entry (PTE).

- The D bit indicates whether the page has been dirtied (i.e., written) since the last time the D bit was cleared.

- The RSW field is reserved for the operating system's use; the hardware ignores it.

- The PPN field holds a physical page number, which is part of a physical address. If this PTE is a leaf, the PPN is part of the translated physical address. Otherwise, the PPN gives the address of the next level of the page table. (Figure 10.16 divides the PPN into two subfields to simplify the description of the address-translation algorithm.)

The other RV64 paging schemes simply add more levels to the page table. Sv48 is nearly identical to Sv39, but its virtual-address space is 2^9 times bigger and its page table is one level deeper.

RV64 supports multiple paging schemes, but we describe only the most popular one, Sv39. Sv39 uses the same 4 KiB base page as Sv32. The page-table entries double in size to eight bytes so they can hold bigger physical addresses. To maintain the invariant that a page table is exactly the size of a page, the radix of the tree correspondingly falls to 2^9. The tree is three levels deep. Sv39's 512 GiB address space is divided into 2^9 *gigapages*, each 1 GiB. Each gigapage is subdivided into 2^9 megapages, which in Sv39 are slightly smaller than in Sv32: 2 MiB. Each megapage is subdivided into 2^9 4 KiB base pages.

Figure 10.17 shows the layout of an Sv39 PTE. It's identical to an Sv32 PTE, except the PPN field has been widened to 44 bits to support 56-bit physical addresses, or 2^{26} GiB of physical-address space.

> ■ *Elaboration: Unused address bits*
>
> Since Sv39's virtual addresses are narrower than an RV64 integer register, you might wonder what becomes of the remaining 25 bits. Sv39 mandates that address bits 63–39 be copies of bit 38. Thus, the valid virtual addresses are $0000\_0000\_0000\_0000_{hex}$– $0000\_003f\_ffff\_ffff_{hex}$ and $ffff\_ffc0\_0000\_0000_{hex}$–$ffff\_ffff\_ffff\_ffff_{hex}$. The gap between these two ranges is, of course, 2^{25} times bigger than the size of the two ranges combined, seemingly wasting 99.999997% of the values a 64-bit register can represent. Why not make better use of those extra 25 bits? The answer is that, as programs grow to require more than 512 GiB of virtual-address space, architects want to increase the address space without breaking backwards compatibility. If we allowed programs to store extra data in the upper 25 bits, it would be impossible to later reclaim those bits to hold bigger addresses. Allowing data storage in unused address bits is a grievous error, but one that has recurred many times in computing history.

An S-mode CSR, `satp` (*Supervisor Address Translation and Protection*), controls the paging system. As Figure 10.18 shows, `satp` has three fields. The MODE field enables paging and selects the page-table depth; Figure 10.19 shows its encoding. The ASID (*Address Space Identifier*) field is optional and can be used to reduce the cost of context switches. Finally, the PPN field holds the physical address of the root page table, divided by the 4 KiB page size. Typically, M-mode software will write zero to `satp` before entering S-mode for the first time, disabling paging, then S-mode software will write it again after setting up the page tables.

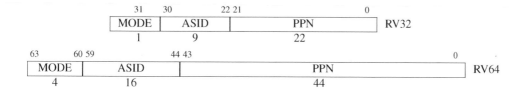

Figure 10.18: The satp CSR. Figures 4.11 and 4.12 of [Waterman and Asanović 2017] are the bases for this figure.

RV32		
Value	Name	Description
0	Bare	No translation or protection.
1	Sv32	Page-based 32-bit virtual addressing.
RV64		
Value	Name	Description
0	Bare	No translation or protection.
8	Sv39	Page-based 39-bit virtual addressing.
9	Sv48	Page-based 48-bit virtual addressing.

Figure 10.19: The encoding of the MODE field in the satp CSR. Table 4.3 of [Waterman and Asanović 2017] is the basis for this figure.

When paging is enabled in the satp register, S-mode and U-mode virtual addresses are translated into physical addresses by traversing the page table, starting at the root. Figure 10.20 depicts this process:

1. satp.PPN gives the base address of the first-level page table, and VA[31:22] gives the first-level index, so the processor reads the PTE located at address (satp.PPN×4096 + VA[31:22]×4).

2. That PTE contains the base address of the second-level page table and VA[21:12] gives the second-level index, so the processor reads the leaf PTE located at (PTE.PPN×4096 + VA[21:12]×4).

3. The leaf PTE's PPN field and the *page offset* (the twelve least-significant bits of the original virtual address) form the final result: the physical address is (LeafPTE.PPN×4096 + VA[11:0]).

The processor then performs the physical memory access. The translation process is almost the same for Sv39 as for Sv32, but with larger PTEs and one more level of indirection. Figure 10.27, at the end of this chapter, gives a complete description of the page-table traversal algorithm, detailing the exceptional conditions and the special case of superpage translations.

That's almost all there is to the RISC-V paging system, save for one wrinkle. If all instruction fetches, loads, and stores resulted in several page-table accesses, then paging would reduce performance substantially! All modern processors reduce this overhead with an address-translation cache (often called a *TLB*, for Translation Lookaside Buffer). To reduce the cost of this cache, most processors don't automatically keep it coherent with the page

Cost

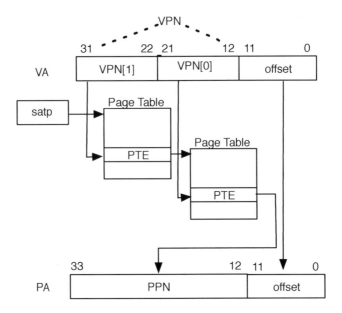

Figure 10.20: Diagram of the Sv32 address-translation process.

table—if the operating system modifies the page table, the cache becomes stale. S-mode adds one more instruction to solve this problem: sfence.vma informs the processor that software may have modified the page tables, so the processor can flush the translation caches accordingly. It takes two optional arguments, which narrow the scope of the cache flush: rs1 indicates which virtual address' translation has been modified in the page table, and rs2 gives the address-space identifier of the process whose page table has been modified. If x0 is given for both, the entire translation cache is flushed.

■ *Elaboration: Address-translation cache coherence in multiprocessors*

sfence.vma only affects the address-translation hardware for the hart that executed the instruction. When a hart changes a page table that another hart is using, the first hart must use an interprocessor interrupt to inform the second hart that it should execute an sfence.vma instruction. This procedure is often referred to as *TLB shootdown*.

10.7 Identification and Performance CSRs

The remaining CSRs identify features of the processor or help measure performance. The identity CSRs are:

- The Machine ISA misa CSR gives the width of the address of the processor (32, 64, or 128 bits) and identifies which instructions extensions are included (Figure 10.21).

- The Vendor ID mvendorid CSR provides the JEDEC manufacturer ID of the provider of the core (Figure 10.22).

XLEN-1 XLEN-2	XLEN-3 26	25 0
MXL[1:0]	0	Extensions[25:0]
2	XLEN-28	26

Figure 10.21: The Machine ISA Register misa **CSR reports the ISA supported. The MXL (Machine XLEN) field encodes the native base integer ISA width: 1 is 32 bits, 2 is 64, and 3 is 128. The Extensions field encodes the presence of the standard extensions, with a single bit per letter of the alphabet (bit 0 encodes presence of extension "A" , bit 1 encodes presence of extension "B", through to bit 25 which encodes "Z").**

- The machine Architecture ID CSR marchid gives the base microarchitecture. Combining mvendorid with marchid uniquely identifies the microarchitecture implemented (Figure 10.23).

- The machine Implementation ID CSR mimpid gives the version of the *implementation* of base microarchitecture in marchid (Figure 10.23).

- The Hart ID CSR mhartid gives the integer ID of the hart currently being run (Figure 10.23).

Here are the measurement CSRs :

- The Machine Time CSR mtime is a 64-bit real time counter (Figure 10.24).

- The Machine Time Compare CSR mtimecmp causes an interrupt when mtime matches or exceeds its value (Figure 10.24).

- The 32-bit machine and supervisor counter-enable CSRs (mcounteren and scounteren) control availability of the hardware performance monitor CSRs at the next-lowest privileged level (Figure 10.25).

- The 32 hardware performance monitor CSRs (mcycle , minstret, mhpmcounter3, ..., mhpmcounter31) count clock cycles, instructions retired, and then up to 29 events selected by the programmer using mhpmevent3, ..., mhpmevent31 CSRs (Figure 10.26).

10.8 Concluding Remarks

Study after study shows that the very best designers produce structures that are faster, smaller, simpler, clearer, and produced with less effort. The differences between the great and the average approach an order of magnitude.

—Fred Brooks, Jr., 1986.

Brooks is a Turing Award laureate and an architect of the IBM System/360 family of computers, which demonstrated the importance of differentiating architecture from implementation. Descendants of that 1964 architecture are still selling today.

The modularity of the RISC-V privileged architectures caters to the needs of a variety of systems. The minimalist Machine mode supports bare-metal embedded applications at low cost. The additional User mode and Physical Memory Protection together enable multitasking in more sophisticated embedded systems. Finally, Supervisor mode and page-based virtual memory provide the flexibility needed to host modern operating systems.

Simplicity

Programmability

XLEN-1		7	6	0
Vendor ID register (mvendorid)			Offset	
XLEN-7			7	

Figure 10.22: The mvendorid **CSR provides the JEDEC manufacturer ID of the core.**

XLEN-1	0
Machine Architecture ID register marchid	
Machine Implementation ID register mimpid	
Machine Hart ID register mhartid	
XLEN	

Figure 10.23: Machine identification CSRs (marchid , mimpid, mhartid) **identify the microarchitecture and implementation of the processor and the number of the currently executing hart thread.**

63	0
Machine Time register mtime	
Machine Time Compare register mtimecmp	
64	

Figure 10.24: Machine time CSRs (mtime and mtimecmp) **measure time and cause an interrupt when** mtime $\geq$ mtimecmp.

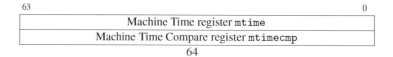

31	30	29	28		6	5	4	3	2	1	0
HPM31	HPM30	HPM29		...		HPM5	HPM4	HPM3	IR	TM	CY
1	1	1		23		1	1	1	1	1	1

Figure 10.25: The counter-enable registers mcounteren **and** scounteren **control the availability of the hardware performance monitoring counters to the next-lowest privileged mode.**

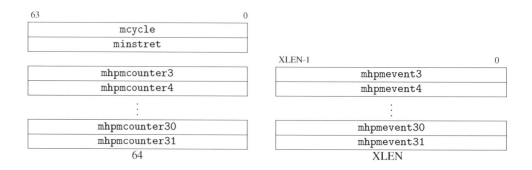

Figure 10.26: Hardware performance monitor CSRs (mcycle , minstret, mhpmcounter3, ..., mhpmcounter31) **and the events they count** mhpmevent3, ..., mhpmevent31. **For RV32 only, reads of the** mcycle, minstret, **and** mhpmcounter*n* **CSRs return the low 32 bits, while reads of the** mcycleh, minstreth, **and** mhpmcounter*n*h **CSRs return bits 63–32 of the corresponding counter.**

1. Let a be satp.*ppn* $\times$ PAGESIZE, and let $i =$ LEVELS $- 1$.

2. Let *pte* be the value of the PTE at address $a + va.vpn[i] \times$ PTESIZE.

3. If $pte.v = 0$, or if $pte.r = 0$ and $pte.w = 1$, stop and raise a page-fault exception.

4. Otherwise, the PTE is valid. If $pte.r = 1$ or $pte.x = 1$, go to step 5. Otherwise, this PTE is a pointer to the next level of the page table. Let $i = i - 1$. If $i < 0$, stop and raise a page-fault exception. Otherwise, let $a = pte.ppn \times$ PAGESIZE and go to step 2.

5. A leaf PTE has been found. Determine if the requested memory access is allowed by the $pte.r$, $pte.w$, $pte.x$, and $pte.u$ bits, given the current privilege mode and the value of the SUM and MXR fields of the mstatus register. If not, stop and raise a page-fault exception.

6. If $i > 0$ and $pa.ppn[i - 1 : 0] \neq 0$, this is a misaligned superpage; stop and raise a page-fault exception.

7. If $pte.a = 0$, or if the memory access is a store and $pte.d = 0$, then either:

   - Raise a page-fault exception, or:

   - Set $pte.a$ to 1 and, if the memory access is a store, also set $pte.d$ to 1.

8. The translation is successful. The translated physical address is given as follows:

   - $pa.pgoff = va.pgoff$.

   - If $i > 0$, then this is a superpage translation and $pa.ppn[i - 1 : 0] = va.vpn[i - 1 : 0]$.

   - $pa.ppn[\text{LEVELS} - 1 : i] = pte.ppn[\text{LEVELS} - 1 : i]$.

Figure 10.27: The complete algorithm for virtual-to-physical address translation. va is the virtual address input and pa is the physical address output. The PAGESIZE constant is 2^{12}. For Sv32, LEVELS=2 and PTESIZE=4, whereas for Sv39, LEVELS=3 and PTESIZE=8. Section 4.3.2 of [Waterman and Asanović 2017] is the basis for this figure.

10.9 To Learn More

A. Waterman and K. Asanović, editors. *The RISC-V Instruction Set Manual Volume II: Privileged Architecture Version 1.10.* May 2017. URL https://riscv.org/specifications/privileged-isa/.

Notes

[1]http://parlab.eecs.berkeley.edu

11 Future RISC-V Optional Extensions

Performance

Performance

Programmability

Fools ignore complexity. Pragmatists suffer it. Some can avoid it. Geniuses remove it.

—Alan Perlis, 1982

The RISC-V Foundation will develop at least eight optional extensions.

11.1 "B" Standard Extension for Bit Manipulation

The B extension offers bit manipulation, including insert, extract, and test bit fields; rotations; funnel shifts; bit and byte permutations; count leading and trailing zeros; and count bits set.

11.2 "E" Standard Extension for Embedded

To reduce the cost of low-end cores, it has 16 fewer registers. RV32E is why the saved and temporary registers are split between the registers 0-15 and 16-31 (Figure 3.2).

11.3 "H" Privileged Architecture Extension for Hypervisor Support

The H extension to the privileged architecture adds a new *hypervisor* mode and a second level of page-based address translation to improve the efficiency of running multiple operating systems on the same machine.

11.4 "J" Standard Extension for Dynamically Translated Languages

Many popular languages are usually implemented via dynamic translation, including Java and Javascript. These languages can benefit from additional ISA support for dynamic checks and garbage collection. (J stands for *Just-In-Time* compiler.)

11.5 "L" Standard Extension for Decimal Floating-Point

The L extension is intended to support decimal floating-point arithmetic as defined in the IEEE 754-2008 standard. The problem with binary numbers is that they cannot represent some common decimal fractions, such as 0.1. The motivation for RV32L is that the computation radix can be identical to the radix of the input and output.

11.6 "N" Standard Extension for User-Level Interrupts

The N extension allows interrupts and exceptions that occur in user-level programs to transfer control directly to a user-level trap handler without invoking the outer execution environment. User-level interrupts are mainly intended to support secure embedded systems with only M-mode and U-mode present (Chapter 10). However, they can also support user-level trap handling in systems running Unix-like operating systems. When used in a Unix environment, conventional signal handling would likely remain, but user-level interrupts could be used as a building block for future extensions that generate user-level events such as garbage collection barriers, integer overflow, and floating-point traps.

Simplicity

Isolation of Arch from Impl

11.7 "P" Standard Extension for Packed-SIMD Instructions

The P extension subdivides the existing architectural registers to provide data-parallel computation on smaller data types. Packed-SIMD designs represent a reasonable design point when reusing existing wide datapath resources. However, if significant additional resources are to be devoted to data-parallel execution, Chapter 8 shows that designs for vector architectures are a better choice, and architects should use the RVV extension.

Performance

11.8 "Q" Standard Extension for Quad-Precision Floating-Point

The Q extension adds 128-bit quad-precision binary floating-point instructions compliant with the IEEE 754-2008 arithmetic standard. The floating-point registers are now extended to hold either a single, double, or quad-precision floating-point value. The quad-precision binary floating-point extension requires RV64IFD.

11.9 Concluding Remarks

> *Simplify, simplify.*
>
> —Henry David Thoreau, an eminent writer of the 19th century, 1854

Having an open, standards-like committee approach to expanding RISC-V hopefully will mean that the feedback and debate will occur *before* the instructions are finalized rather than afterwards, when it's too late to change. In the ideal case, a few members will implement the proposal before it is ratified, which FPGAs make much easier to do. Proposing instruction extensions via the RISC-V Foundation committees will also be a fair amount of work, which will keep the rate of change slow, unlike what happened to x86-32 (see Figure 1.2 on page 3 in Chapter 1). Don't forget that everything in this chapter will be optional, despite how many extensions are adopted.

Our hope is that RISC-V can evolve with technological demands while maintaining its reputation as a simple, efficient ISA. If it succeeds, RISC-V will we be a significant break from the incremental ISAs of the past.

Room for Growth

Elegance

A RISC-V Instruction Listings

Simplicity is the keynote of all true elegance.

—Coco Chanel, 1923

This appendix lists all the instructions for RV32/64I, all the extensions covered in this book except RVV (RVM, RVA, RVF, RVD, and RVC), and all the pseudoinstructions. Each entry has the instruction name, operands, a register-transfer level definition, instruction format type, English description, compressed versions (if any), and a figure showing the actual layout with opcodes. We think you have everything you need to understand all the instructions in these compact summaries. However, if you want even more detail, refer to the official RISC-V specifications [Waterman and Asanović 2017].

To help readers find the desired instruction in this appendix, the header of the left (even) page contains the first instruction from the top of that page and the header on the right (odd) page contains the last instruction from at the bottom of that page. The format is similar to the headers of dictionaries, which helps you search for the page that your word is on. For example, the header of the next *even* page shows **AMOADD.W**, the first instruction on the page, and the header of the following odd page shows **AMOMINU.D**, the last instruction on that page. These are the two pages where you would find any of these 10 instructions: `amoadd.w`, `amoand.d`, `amoand.w`, `amomax.d`, `amomax.w`, `amomaxu.d`, `amomaxu.w`, `amomin.d`, `amomin.w`, and `amominu.d`.

add rd, rs1, rs2   x[rd] = x[rs1] + x[rs2]

Add. R-type, RV32I and RV64I.

Adds register x[*rs2*] to register x[*rs1*] and writes the result to x[*rd*]. Arithmetic overflow is ignored.

Compressed forms: **c.add** rd, rs2; **c.mv** rd, rs2

31	25	24	20	19	15	14	12	11	7	6	0
0000000		rs2		rs1		000		rd		0110011	

---

addi rd, rs1, immediate   x[rd] = x[rs1] + sext(immediate)

Add Immediate. I-type, RV32I and RV64I.

Adds the sign-extended *immediate* to register x[*rs1*] and writes the result to x[*rd*]. Arithmetic overflow is ignored.

Compressed forms: **c.li** rd, imm; **c.addi** rd, imm; **c.addi16sp** imm; **c.addi4spn** rd, imm

31	20	19	15	14	12	11	7	6	0
immediate[11:0]		rs1		000		rd		0010011	

---

addiw rd, rs1, immediate x[rd] = sext((x[rs1] + sext(immediate))[31:0])

Add Word Immediate. I-type, RV64I only.

Adds the sign-extended *immediate* to x[*rs1*], truncates the result to 32 bits, and writes the sign-extended result to x[*rd*]. Arithmetic overflow is ignored.

Compressed form: **c.addiw** rd, imm

31	20	19	15	14	12	11	7	6	0
immediate[11:0]		rs1		000		rd		0011011	

---

addw rd, rs1, rs2   x[rd] = sext((x[rs1] + x[rs2])[31:0])

Add Word. R-type, RV64I only.

Adds register x[*rs2*] to register x[*rs1*], truncates the result to 32 bits, and writes the sign-extended result to x[*rd*]. Arithmetic overflow is ignored.

Compressed form: **c.addw** rd, rs2

31	25	24	20	19	15	14	12	11	7	6	0
0000000		rs2		rs1		000		rd		0111011	

---

amoadd.d rd, rs2, (rs1)   x[rd] = AMO64(M[x[rs1]] + x[rs2])

Atomic Memory Operation: Add Doubleword. R-type, RV64A only.

Atomically, let *t* be the value of the memory doubleword at address x[*rs1*], then set that memory doubleword to *t* + x[*rs2*]. Set x[*rd*] to *t*.

31	27	26	25	24	20	19	15	14	12	11	7	6	0
00000		aq	rl	rs2		rs1		011		rd		0101111	

amoadd.w rd, rs2, (rs1)   `x[rd] = AMO32(M[x[rs1]] + x[rs2])`

Atomic Memory Operation: Add Word. R-type, RV32A and RV64A.

Atomically, let t be the value of the memory word at address x[*rs1*], then set that memory word to t + x[*rs2*]. Set x[*rd*] to the sign extension of t.

31	27 26	25 24		20 19	15 14	12 11	7 6	0
00000	aq	rl	rs2	rs1	010	rd	0101111

amoand.d rd, rs2, (rs1)   `x[rd] = AMO64(M[x[rs1]] & x[rs2])`

Atomic Memory Operation: AND Doubleword. R-type, RV64A only.

Atomically, let t be the value of the memory doubleword at address x[*rs1*], then set that memory doubleword to the bitwise AND of t and x[*rs2*]. Set x[*rd*] to t.

31	27 26	25 24		20 19	15 14	12 11	7 6	0
01100	aq	rl	rs2	rs1	011	rd	0101111

amoand.w rd, rs2, (rs1)   `x[rd] = AMO32(M[x[rs1]] & x[rs2])`

Atomic Memory Operation: AND Word. R-type, RV32A and RV64A.

Atomically, let t be the value of the memory word at address x[*rs1*], then set that memory word to the bitwise AND of t and x[*rs2*]. Set x[*rd*] to the sign extension of t.

31	27 26	25 24		20 19	15 14	12 11	7 6	0
01100	aq	rl	rs2	rs1	010	rd	0101111

amomax.d rd, rs2, (rs1)   `x[rd] = AMO64(M[x[rs1]] MAX x[rs2])`

Atomic Memory Operation: Maximum Doubleword. R-type, RV64A only.

Atomically, let t be the value of the memory doubleword at address x[*rs1*], then set that memory doubleword to the larger of t and x[*rs2*], using a two's complement comparison. Set x[*rd*] to t.

31	27 26	25 24		20 19	15 14	12 11	7 6	0
10100	aq	rl	rs2	rs1	011	rd	0101111

amomax.w rd, rs2, (rs1)   `x[rd] = AMO32(M[x[rs1]] MAX x[rs2])`

Atomic Memory Operation: Maximum Word. R-type, RV32A and RV64A.

Atomically, let t be the value of the memory word at address x[*rs1*], then set that memory word to the larger of t and x[*rs2*], using a two's complement comparison. Set x[*rd*] to the sign extension of t.

31	27 26	25 24		20 19	15 14	12 11	7 6	0
10100	aq	rl	rs2	rs1	010	rd	0101111

amomaxu.d rd, rs2, (rs1)   x[rd] = AMO64(M[x[rs1]] MAXU x[rs2])

Atomic Memory Operation: Maximum Doubleword, Unsigned. R-type, RV64A only.

Atomically, let t be the value of the memory doubleword at address x[*rs1*], then set that memory doubleword to the larger of t and x[*rs2*], using an unsigned comparison. Set x[*rd*] to t.

31	27	26	25	24	20	19	15	14	12	11	7	6	0
11100		aq	rl	rs2		rs1		011		rd		0101111	

amomaxu.w rd, rs2, (rs1)   x[rd] = AMO32(M[x[rs1]] MAXU x[rs2])

Atomic Memory Operation: Maximum Word, Unsigned. R-type, RV32A and RV64A.

Atomically, let t be the value of the memory word at address x[*rs1*], then set that memory word to the larger of t and x[*rs2*], using an unsigned comparison. Set x[*rd*] to the sign extension of t.

31	27	26	25	24	20	19	15	14	12	11	7	6	0
11100		aq	rl	rs2		rs1		010		rd		0101111	

amomin.d rd, rs2, (rs1)   x[rd] = AMO64(M[x[rs1]] MIN x[rs2])

Atomic Memory Operation: Minimum Doubleword. R-type, RV64A only.

Atomically, let t be the value of the memory doubleword at address x[*rs1*], then set that memory doubleword to the smaller of t and x[*rs2*], using a two's complement comparison. Set x[*rd*] to t.

31	27	26	25	24	20	19	15	14	12	11	7	6	0
10000		aq	rl	rs2		rs1		011		rd		0101111	

amomin.w rd, rs2, (rs1)   x[rd] = AMO32(M[x[rs1]] MIN x[rs2])

Atomic Memory Operation: Minimum Word. R-type, RV32A and RV64A.

Atomically, let t be the value of the memory word at address x[*rs1*], then set that memory word to the smaller of t and x[*rs2*], using a two's complement comparison. Set x[*rd*] to the sign extension of t.

31	27	26	25	24	20	19	15	14	12	11	7	6	0
10000		aq	rl	rs2		rs1		010		rd		0101111	

amominu.d rd, rs2, (rs1)   x[rd] = AMO64(M[x[rs1]] MINU x[rs2])

Atomic Memory Operation: Minimum Doubleword, Unsigned. R-type, RV64A only.

Atomically, let t be the value of the memory doubleword at address x[*rs1*], then set that memory doubleword to the smaller of t and x[*rs2*], using an unsigned comparison. Set x[*rd*] to t.

31	27	26	25	24	20	19	15	14	12	11	7	6	0
11000		aq	rl	rs2		rs1		011		rd		0101111	

amominu.w rd, rs2, (rs1)   x[rd] = AMO32(M[x[rs1]] MINU x[rs2])

Atomic Memory Operation: Minimum Word, Unsigned. R-type, RV32A and RV64A.

Atomically, let t be the value of the memory word at address x[*rs1*], then set that memory word to the smaller of t and x[*rs2*], using an unsigned comparison. Set x[*rd*] to the sign extension of t.

31	27 26	25	24	20 19	15 14	12 11	7 6	0
11000	aq	rl	rs2	rs1	010	rd	0101111	

amoor.d rd, rs2, (rs1)   x[rd] = AMO64(M[x[rs1]] | x[rs2])

Atomic Memory Operation: OR Doubleword. R-type, RV64A only.

Atomically, let t be the value of the memory doubleword at address x[*rs1*], then set that memory doubleword to the bitwise OR of t and x[*rs2*]. Set x[*rd*] to t.

31	27 26	25	24	20 19	15 14	12 11	7 6	0
01000	aq	rl	rs2	rs1	011	rd	0101111	

amoor.w rd, rs2, (rs1)   x[rd] = AMO32(M[x[rs1]] | x[rs2])

Atomic Memory Operation: OR Word. R-type, RV32A and RV64A.

Atomically, let t be the value of the memory word at address x[*rs1*], then set that memory word to the bitwise OR of t and x[*rs2*]. Set x[*rd*] to the sign extension of t.

31	27 26	25	24	20 19	15 14	12 11	7 6	0
01000	aq	rl	rs2	rs1	010	rd	0101111	

amoswap.d rd, rs2, (rs1)   x[rd] = AMO64(M[x[rs1]] SWAP x[rs2])

Atomic Memory Operation: Swap Doubleword. R-type, RV64A only.

Atomically, let t be the value of the memory doubleword at address x[*rs1*], then set that memory doubleword to x[*rs2*]. Set x[*rd*] to t.

31	27 26	25	24	20 19	15 14	12 11	7 6	0
00001	aq	rl	rs2	rs1	011	rd	0101111	

amoswap.w rd, rs2, (rs1)   x[rd] = AMO32(M[x[rs1]] SWAP x[rs2])

Atomic Memory Operation: Swap Word. R-type, RV32A and RV64A.

Atomically, let t be the value of the memory word at address x[*rs1*], then set that memory word to x[*rs2*]. Set x[*rd*] to the sign extension of t.

31	27 26	25	24	20 19	15 14	12 11	7 6	0
00001	aq	rl	rs2	rs1	010	rd	0101111	

amoxor.d rd, rs2, (rs1) x[rd] = AMO64(M[x[rs1]] ^ x[rs2])

Atomic Memory Operation: XOR Doubleword. R-type, RV64A only.
Atomically, let *t* be the value of the memory doubleword at address x[*rs1*], then set that
memory doubleword to the bitwise XOR of *t* and x[*rs2*]. Set x[*rd*] to *t*.

31		27 26	25 24		20 19		15 14	12 11		7 6		0
00100		aq	rl	rs2		rs1		011	rd		0101111	

amoxor.w rd, rs2, (rs1) x[rd] = AMO32(M[x[rs1]] ^ x[rs2])

Atomic Memory Operation: XOR Word. R-type, RV32A and RV64A.
Atomically, let *t* be the value of the memory word at address x[*rs1*], then set that memory
word to the bitwise XOR of *t* and x[*rs2*]. Set x[*rd*] to the sign extension of *t*.

31		27 26	25 24		20 19		15 14	12 11		7 6		0
00100		aq	rl	rs2		rs1		010	rd		0101111	

and rd, rs1, rs2 x[rd] = x[rs1] & x[rs2]

AND. R-type, RV32I and RV64I.
Computes the bitwise AND of registers x[*rs1*] and x[*rs2*] and writes the result to x[*rd*].
Compressed form: **c.and** rd, rs2

31		25 24		20 19		15 14	12 11		7 6		0
0000000		rs2		rs1		111	rd		0110011	

andi rd, rs1, immediate x[rd] = x[rs1] & sext(immediate)

AND Immediate. I-type, RV32I and RV64I.
Computes the bitwise AND of the sign-extended *immediate* and register x[*rs1*] and writes the
result to x[*rd*].
Compressed form: **c.andi** rd, imm

31		20 19		15 14	12 11		7 6		0
immediate[11:0]		rs1		111	rd		0010011	

auipc rd, immediate x[rd] = pc + sext(immediate[31:12] << 12)

Add Upper Immediate to PC. U-type, RV32I and RV64I.
Adds the sign-extended 20-bit *immediate*, left-shifted by 12 bits, to the *pc*, and writes the
result to x[*rd*].

31		12 11		7 6		0
immediate[31:12]		rd		0010111	

beq rs1, rs2, offset   if (rs1 == rs2) pc += sext(offset)

Branch if Equal. B-type, RV32I and RV64I.

If register x[*rs1*] equals register x[*rs2*], set the *pc* to the current *pc* plus the sign-extended *offset*.

Compressed form: **c.beqz** rs1, offset

31	25	24	20	19	15	14	12	11	7	6	0
offset[12|10:5]		rs2		rs1		000		offset[4:1|11]		1100011	

beqz rs1, offset   if (rs1 == 0) pc += sext(offset)

Branch if Equal to Zero. Pseudoinstruction, RV32I and RV64I.

Expands to **beq** rs1, x0, offset.

bge rs1, rs2, offset   if (rs1 $\geq_s$ rs2) pc += sext(offset)

Branch if Greater Than or Equal. B-type, RV32I and RV64I.

If register x[*rs1*] is at least x[*rs2*], treating the values as two's complement numbers, set the *pc* to the current *pc* plus the sign-extended *offset*.

31	25	24	20	19	15	14	12	11	7	6	0
offset[12|10:5]		rs2		rs1		101		offset[4:1|11]		1100011	

bgeu rs1, rs2, offset   if (rs1 $\geq_u$ rs2) pc += sext(offset)

Branch if Greater Than or Equal, Unsigned. B-type, RV32I and RV64I.

If register x[*rs1*] is at least x[*rs2*], treating the values as unsigned numbers, set the *pc* to the current *pc* plus the sign-extended *offset*.

31	25	24	20	19	15	14	12	11	7	6	0
offset[12|10:5]		rs2		rs1		111		offset[4:1|11]		1100011	

bgez rs1, offset   if (rs1 $\geq_s$ 0) pc += sext(offset)

Branch if Greater Than or Equal to Zero. Pseudoinstruction, RV32I and RV64I.

Expands to **bge** rs1, x0, offset.

bgt rs1, rs2, offset   if (rs1 $>_s$ rs2) pc += sext(offset)

Branch if Greater Than. Pseudoinstruction, RV32I and RV64I.

Expands to **blt** rs2, rs1, offset.

bgtu rs1, rs2, offset   if (rs1 $>_u$ rs2) pc += sext(offset)

Branch if Greater Than, Unsigned. Pseudoinstruction, RV32I and RV64I.

Expands to **bltu** rs2, rs1, offset.

bgtz rs2, offset   if (rs2 $>_s$ 0) pc += sext(offset)
Branch if Greater Than Zero. Pseudoinstruction, RV32I and RV64I.
Expands to **blt** x0, rs2, offset.

---

ble rs1, rs2, offset   if (rs1 $\leq_s$ rs2) pc += sext(offset)
Branch if Less Than or Equal. Pseudoinstruction, RV32I and RV64I.
Expands to **bge** rs2, rs1, offset.

---

bleu rs1, rs2, offset   if (rs1 $\leq_u$ rs2) pc += sext(offset)
Branch if Less Than or Equal, Unsigned. Pseudoinstruction, RV32I and RV64I.
Expands to **bgeu** rs2, rs1, offset.

---

blez rs2, offset   if (rs2 $\leq_s$ 0) pc += sext(offset)
Branch if Less Than or Equal to Zero. Pseudoinstruction, RV32I and RV64I.
Expands to **bge** x0, rs2, offset.

---

blt rs1, rs2, offset   if (rs1 $<_s$ rs2) pc += sext(offset)
Branch if Less Than. B-type, RV32I and RV64I.
If register x[*rs1*] is less than x[*rs2*], treating the values as two's complement numbers, set the
pc to the current *pc* plus the sign-extended *offset*.

31 25	24 20	19 15	14 12	11 7	6 0
offset[12|10:5]	rs2	rs1	100	offset[4:1|11]	1100011

---

bltz rs1, offset   if (rs1 $<_s$ 0) pc += sext(offset)
Branch if Less Than Zero. Pseudoinstruction, RV32I and RV64I.
Expands to **blt** rs1, x0, offset.

---

bltu rs1, rs2, offset   if (rs1 $<_u$ rs2) pc += sext(offset)
Branch if Less Than, Unsigned. B-type, RV32I and RV64I.
If register x[*rs1*] is less than x[*rs2*], treating the values as unsigned numbers, set the *pc* to the
current *pc* plus the sign-extended *offset*.

31 25	24 20	19 15	14 12	11 7	6 0
offset[12|10:5]	rs2	rs1	110	offset[4:1|11]	1100011

bne rs1, rs2, offset   if (rs1 $\neq$ rs2) pc += sext(offset)

Branch if Not Equal. B-type, RV32I and RV64I.

If register x[*rs1*] does not equal register x[*rs2*], set the *pc* to the current *pc* plus the sign-extended *offset*.

Compressed form: **c.bnez** rs1, offset

31	25 24	20 19	15 14	12 11	7 6	0
offset[12|10:5]	rs2	rs1	001	offset[4:1|11]	1100011	

bnez rs1, offset   if (rs1 $\neq$ 0) pc += sext(offset)

Branch if Not Equal to Zero. Pseudoinstruction, RV32I and RV64I.

Expands to **bne** rs1, x0, offset.

c.add rd, rs2   x[rd] = x[rd] + x[rs2]

Add. RV32IC and RV64IC.

Expands to **add** rd, rd, rs2. Invalid when rd=x0 or rs2=x0.

15	13	12	11	7 6	2 1	0
100		1	rd	rs2	10	

c.addi rd, imm   x[rd] = x[rd] + sext(imm)

Add Immediate. RV32IC and RV64IC.

Expands to **addi** rd, rd, imm.

15	13	12	11	7 6	2 1	0
000		imm[5]	rd	imm[4:0]	01	

c.addi16sp imm   x[2] = x[2] + sext(imm)

Add Immediate, Scaled by 16, to Stack Pointer. RV32IC and RV64IC.

Expands to **addi** x2, x2, imm. Invalid when imm=0.

15	13	12	11	7 6	2 1	0
011		imm[9]	00010	imm[4|6|8:7|5]	01	

c.addi4spn rd′, uimm   x[8+rd′] = x[2] + uimm

Add Immediate, Scaled by 4, to Stack Pointer, Nondestructive. RV32IC and RV64IC.

Expands to **addi** rd, x2, uimm, where rd=8+rd′. Invalid when uimm=0.

15	13 12	5 4	2 1	0
000	uimm[5:4|9:6|2|3]	rd′	00	

c.addiw rd, imm   x[rd] = sext((x[rd] + sext(imm))[31:0])

Add Word Immediate. RV64IC only.

Expands to **addiw** rd, rd, imm. Invalid when rd=x0.

15	13	12	11		7 6		2 1	0
	001	imm[5]		rd		imm[4:0]		01

---

c.and rd′, rs2′   x[8+rd′] = x[8+rd′] & x[8+rs2′]

AND. RV32IC and RV64IC.

Expands to **and** rd, rd, rs2, where rd=8+rd′ and rs2=8+rs2′.

15		10 9		7 6	5 4		2 1	0
	100011		rd′		11	rs2′		01

---

c.addw rd′, rs2′   x[8+rd′] = sext((x[8+rd′] + x[8+rs2′])[31:0])

Add Word. RV64IC only.

Expands to **addw** rd, rd, rs2, where rd=8+rd′ and rs2=8+rs2′.

15		10 9		7 6	5 4		2 1	0
	100111		rd′		01	rs2′		01

---

c.andi rd′, imm   x[8+rd′] = x[8+rd′] & sext(imm)

AND Immediate. RV32IC and RV64IC.

Expands to **andi** rd, rd, imm, where rd=8+rd′.

15	13	12	11 10 9		7 6		2 1	0
	100	imm[5]	10	rd′		imm[4:0]		01

---

c.beqz rs1′, offset   if (x[8+rs1′] == 0) pc += sext(offset)

Branch if Equal to Zero. RV32IC and RV64IC.

Expands to **beq** rs1, x0, offset, where rs1=8+rs1′.

15	13 12		10 9		7 6		2 1	0
	110	offset[8|4:3]		rs1′		offset[7:6|2:1|5]		01

---

c.bnez rs1′, offset   if (x[8+rs1′] ≠ 0) pc += sext(offset)

Branch if Not Equal to Zero. RV32IC and RV64IC.

Expands to **bne** rs1, x0, offset, where rs1=8+rs1′.

15	13 12		10 9		7 6		2 1	0
	111	offset[8|4:3]		rs1′		offset[7:6|2:1|5]		01

c.ebreak   RaiseException(Breakpoint)

Environment Breakpoint. RV31IC and RV64IC.
Expands to **ebreak**.

15 13	12	11 7	6 2	1 0
100	1	00000	00000	10

c.fld rd', uimm(rs1')   f[8+rd'] = M[x[8+rs1'] + uimm][63:0]

Floating-point Load Doubleword. RV32DC and RV64DC.
Expands to **fld** rd, uimm(rs1), where rd=8+rd' and rs1=8+rs1'.

15 13	12 10	9 7	6 5	4 2	1 0
001	uimm[5:3]	rs1'	uimm[7:6]	rd'	00

c.fldsp rd, uimm(x2)   f[rd] = M[x[2] + uimm][63:0]

Floating-point Load Doubleword, Stack-Pointer Relative. RV32DC and RV64DC.
Expands to **fld** rd, uimm(x2).

15 13	12	11 7	6 2	1 0
001	uimm[5]	rd	uimm[4:3|8:6]	10

c.flw rd', uimm(rs1')   f[8+rd'] = M[x[8+rs1'] + uimm][31:0]

Floating-point Load Word. RV32FC only.
Expands to **flw** rd, uimm(rs1), where rd=8+rd' and rs1=8+rs1'.

15 13	12 10	9 7	6 5	4 2	1 0
011	uimm[5:3]	rs1'	uimm[2|6]	rd'	00

c.flwsp rd, uimm(x2)   f[rd] = M[x[2] + uimm][31:0]

Floating-point Load Word, Stack-Pointer Relative. RV32FC only.
Expands to **flw** rd, uimm(x2).

15 13	12	11 7	6 2	1 0
011	uimm[5]	rd	uimm[4:2|7:6]	10

c.fsd rs2', uimm(rs1')   M[x[8+rs1'] + uimm][63:0] = f[8+rs2']

Floating-point Store Doubleword. RV32DC and RV64DC.
Expands to **fsd** rs2, uimm(rs1), where rs2=8+rs2' and rs1=8+rs1'.

15 13	12 10	9 7	6 5	4 2	1 0
101	uimm[5:3]	rs1'	uimm[7:6]	rs2'	00

c.fsdsp rs2, uimm(x2)
M[x[2] + uimm][63:0] = f[rs2]

Floating-point Store Doubleword, Stack-Pointer Relative. RV32DC and RV64DC.
Expands to **fsd** rs2, uimm(x2).

15 13	12 7	6 2	1 0
101	uimm[5:3|8:6]	rs2	10

c.fsw rs2′, uimm(rs1′)
M[x[8+rs1′] + uimm][31:0] = f[8+rs2′]

Floating-point Store Word. RV32FC only.
Expands to **fsw** rs2, uimm(rs1), where rs2=8+rs2′ and rs1=8+rs1′.

15 13	12 10	9 7	6 5	4 2	1 0
111	uimm[5:3]	rs1′	uimm[2|6]	rs2′	00

c.fswsp rs2, uimm(x2)
M[x[2] + uimm][31:0] = f[rs2]

Floating-point Store Word, Stack-Pointer Relative. RV32FC only.
Expands to **fsw** rs2, uimm(x2).

15 13	12 7	6 2	1 0
111	uimm[5:2|7:6]	rs2	10

c.j offset
pc += sext(offset)

Jump. RV32IC and RV64IC.
Expands to **jal** x0, offset.

15 13	12 2	1 0
101	offset[11|4|9:8|10|6|7|3:1|5]	01

# c.jal offset
x[1] = pc+2; pc += sext(offset)

Jump and Link. RV32IC only.
Expands to **jal** x1, offset.

15 13	12 2	1 0
001	offset[11|4|9:8|10|6|7|3:1|5]	01

# c.jalr rs1
t = pc+2; pc = x[rs1]; x[1] = t

Jump and Link Register. RV32IC and RV64IC.
Expands to **jalr** x1, 0(rs1). Invalid when rs1=x0.

15 13	12	11 7	6 2	1 0
100	1	rs1	00000	10

c.jr rs1   pc = x[rs1]

Jump Register. RV32IC and RV64IC.
Expands to **jalr** x0, 0(rs1). Invalid when rs1=x0.

15 13	12	11 rs1 7 6	00000	2 1 0
100	0	rs1	00000	10

---

c.ld rd', uimm(rs1')   x[8+rd'] = M[x[8+rs1'] + uimm][63:0]

Load Doubleword. RV64IC only.
Expands to **ld** rd, uimm(rs1), where rd=8+rd' and rs1=8+rs1'.

15 13	12 10 9	7 6	5 4	2 1 0
011	uimm[5:3]	rs1'| uimm[7:6]	rd'	00

---

c.ldsp rd, uimm(x2)   x[rd] = M[x[2] + uimm][63:0]

Load Doubleword, Stack-Pointer Relative. RV64IC only.
Expands to **ld** rd, uimm(x2). Invalid when rd=x0.

15 13	12	11 rd 7 6	2 1 0
011	uimm[5]	rd	uimm[4:3|8:6]	10

---

c.li rd, imm   x[rd] = sext(imm)

Load Immediate. RV32IC and RV64IC.
Expands to **addi** rd, x0, imm.

15 13	12	11 rd 7 6	imm[4:0]	2 1 0
010	imm[5]	rd	imm[4:0]	01

---

c.lui rd, imm   x[rd] = sext(imm[17:12] << 12)

Load Upper Immediate. RV32IC and RV64IC.
Expands to **lui** rd, imm. Invalid when rd=x2 or imm=0.

15 13	12	11 rd 7 6	imm[16:12]	2 1 0
011	imm[17]	rd	imm[16:12]	01

---

c.lw rd', uimm(rs1')   x[8+rd'] = sext(M[x[8+rs1'] + uimm][31:0])

Load Word. RV32IC and RV64IC.
Expands to **lw** rd, uimm(rs1), where rd=8+rd' and rs1=8+rs1'.

15 13	12 10 9	7 6	5 4	2 1 0
010	uimm[5:3]	rs1'| uimm[2|6]	rd'	00

c.lwsp rd, uimm(x2)   x[rd] = sext(M[x[2] + uimm][31:0])
Load Word, Stack-Pointer Relative. RV32IC and RV64IC.
Expands to **lw** rd, uimm(x2). Invalid when rd=x0.

15 13	12	11 7	6 2	1 0
010	uimm[5]	rd	uimm[4:2|7:6]	10

---

c.mv rd, rs2   x[rd] = x[rs2]
Move. RV32IC and RV64IC.
Expands to **add** rd, x0, rs2. Invalid when rs2=x0.

15 13	12	11 7	6 2	1 0
100	0	rd	rs2	10

---

c.or rd', rs2'   x[8+rd'] = x[8+rd'] | x[8+rs2']
OR. RV32IC and RV64IC.
Expands to **or** rd, rd, rs2, where rd=8+rd' and rs2=8+rs2'.

15 10	9 7	6 5	4 2	1 0
100011	rd'	10	rs2'	01

---

c.sd rs2', uimm(rs1')   M[x[8+rs1'] + uimm][63:0] = x[8+rs2']
Store Doubleword. RV64IC only.
Expands to **sd** rs2, uimm(rs1), where rs2=8+rs2' and rs1=8+rs1'.

15 13	12 10	9 7	6 5	4 2	1 0
111	uimm[5:3]	rs1'	uimm[7:6]	rs2'	00

---

c.sdsp rs2, uimm(x2)   M[x[2] + uimm][63:0] = x[rs2]
Store Doubleword, Stack-Pointer Relative. RV64IC only.
Expands to **sd** rs2, uimm(x2).

15 13	12 7	6 2	1 0
111	uimm[5:3|8:6]	rs2	10

---

c.slli rd, uimm   x[rd] = x[rd] << uimm
Shift Left Logical Immediate. RV32IC and RV64IC.
Expands to **slli** rd, rd, uimm.

15 13	12	11 7	6 2	1 0
000	uimm[5]	rd	uimm[4:0]	10

c.srai rd', uimm $x[8+rd'] = x[8+rd'] >>_s uimm$

Shift Right Arithmetic Immediate. RV32IC and RV64IC.

Expands to **srai** rd, rd, uimm, where rd=8+rd'.

15	13 12	11 10 9	7 6	2 1	0
100	uimm[5] 01	rd'	uimm[4:0]	01	

c.srli rd', uimm $x[8+rd'] = x[8+rd'] >>_u uimm$

Shift Right Logical Immediate. RV32IC and RV64IC.

Expands to **srli** rd, rd, uimm, where rd=8+rd'.

15	13 12	11 10 9	7 6	2 1	0
100	uimm[5] 00	rd'	uimm[4:0]	01	

c.sub rd', rs2' $x[8+rd'] = x[8+rd'] - x[8+rs2']$

Subtract. RV32IC and RV64IC.

Expands to **sub** rd, rd, rs2, where rd=8+rd' and rs2=8+rs2'.

15	10 9	7 6	5 4	2 1	0
100011	rd'	00	rs2'	01	

c.subw rd', rs2' $x[8+rd'] = sext((x[8+rd'] - x[8+rs2'])[31:0])$

Subtract Word. RV64IC only.

Expands to **subw** rd, rd, rs2, where rd=8+rd' and rs2=8+rs2'.

15	10 9	7 6	5 4	2 1	0
100111	rd'	00	rs2'	01	

c.sw rs2', uimm(rs1') $M[x[8+rs1'] + uimm][31:0] = x[8+rs2']$

Store Word. RV32IC and RV64IC.

Expands to **sw** rs2, uimm(rs1), where rs2=8+rs2' and rs1=8+rs1'.

15	13 12	10 9	7 6	5 4	2 1	0
110	uimm[5:3]	rs1'	uimm[2|6]	rs2'	00	

c.swsp rs2, uimm(x2) $M[x[2] + uimm][31:0] = x[rs2]$

Store Word, Stack-Pointer Relative. RV32IC and RV64IC.

Expands to **sw** rs2, uimm(x2).

15	13 12	7 6	2 1	0
110	uimm[5:2|7:6]	rs2	10	

c.xor rd', rs2'

x[8+rd'] = x[8+rd'] ^ x[8+rs2']

Exclusive-OR. RV32IC and RV64IC.
Expands to **xor** rd, rd, rs2, where rd=8+rd' and rs2=8+rs2'.

15	10 9	7 6	5 4	2 1	0
100011	rd'	01	rs2'	01	

call rd, symbol

x[rd] = pc+8; pc = &symbol

Call. Pseudoinstruction, RV32I and RV64I.
Writes the address of the next instruction (*pc+8*) to x[*rd*], then sets the *pc* to *symbol*. Expands to **auipc** rd, offsetHi then **jalr** rd, offsetLo(rd). If *rd* is omitted, x1 is implied.

csrr rd, csr

x[rd] = CSRs[csr]

Control and Status Register Read. Pseudoinstruction, RV32I and RV64I.
Copies control and status register *csr* to x[*rd*]. Expands to **csrrs** rd, csr, x0.

csrc csr, rs1

CSRs[csr] &= ~x[rs1]

Control and Status Register Clear. Pseudoinstruction, RV32I and RV64I.
For each bit set in x[*rs1*], clear the corresponding bit in control and status register *csr*. Expands to **csrrc** x0, csr, rs1.

csrci csr, zimm[4:0]

CSRs[csr] &= ~zimm

Control and Status Register Clear Immediate. Pseudoinstruction, RV32I and RV64I.
For each bit set in the five-bit zero-extended immediate, clear the corresponding bit in control and status register *csr*. Expands to **csrrci** x0, csr, zimm.

csrrc rd, csr, rs1

t = CSRs[csr]; CSRs[csr] = t&~x[rs1]; x[rd] = t

Control and Status Register Read and Clear. I-type, RV32I and RV64I.
Let *t* be the value of control and status register *csr*. Write the bitwise AND of *t* and the ones' complement of x[*rs1*] to the *csr*, then write *t* to x[*rd*].

31	20 19	15 14	12 11	7 6	0
csr	rs1	011	rd	1110011	

csrrci rd, csr, zimm[4:0] t = CSRs[csr]; CSRs[csr] = t &~zimm; x[rd] = t

Control and Status Register Read and Clear Immediate. I-type, RV32I and RV64I.

Let t be the value of control and status register *csr*. Write the bitwise AND of t and the ones' complement of the five-bit zero-extended immediate *zimm* to the *csr*, then write t to x[*rd*]. (Bits 5 and above in the *csr* are not modified.)

31	20 19	15 14	12 11	7 6	0
csr	zimm[4:0]	111	rd	1110011	

csrrs rd, csr, rs1 t = CSRs[csr]; CSRs[csr] = t | x[rs1]; x[rd] = t

Control and Status Register Read and Set. I-type, RV32I and RV64I.

Let t be the value of control and status register *csr*. Write the bitwise OR of t and x[*rs1*] to the *csr*, then write t to x[*rd*].

31	20 19	15 14	12 11	7 6	0
csr	rs1	010	rd	1110011	

csrrsi rd, csr, zimm[4:0] t = CSRs[csr]; CSRs[csr] = t | zimm; x[rd] = t

Control and Status Register Read and Set Immediate. I-type, RV32I and RV64I.

Let t be the value of control and status register *csr*. Write the bitwise OR of t and the five-bit zero-extended immediate *zimm* to the *csr*, then write t to x[*rd*]. (Bits 5 and above in the *csr* are not modified.)

31	20 19	15 14	12 11	7 6	0
csr	zimm[4:0]	110	rd	1110011	

csrrw rd, csr, rs1 t = CSRs[csr]; CSRs[csr] = x[rs1]; x[rd] = t

Control and Status Register Read and Write. I-type, RV32I and RV64I.

Let t be the value of control and status register *csr*. Copy x[*rs1*] to the *csr*, then write t to x[*rd*].

31	20 19	15 14	12 11	7 6	0
csr	rs1	001	rd	1110011	

csrrwi rd, csr, zimm[4:0]   x[rd] = CSRs[csr]; CSRs[csr] = zimm

Control and Status Register Read and Write Immediate. I-type, RV32I and RV64I.

Copies the control and status register *csr* to x[*rd*], then writes the five-bit zero-extended immediate *zimm* to the *csr*.

31	20 19	15 14	12 11	7 6	0
csr	zimm[4:0]	101	rd	1110011	

csrs csr, rs1 　　　　　　　　　　CSRs[csr] |= x[rs1]

Control and Status Register Set. Pseudoinstruction, RV32I and RV64I.

For each bit set in x[*rs1*], set the corresponding bit in control and status register *csr*. Expands to **csrrs** x0, csr, rs1.

---

csrsi csr, zimm[4:0] 　　　　　　　　　CSRs[csr] |= zimm

Control and Status Register Set Immediate. Pseudoinstruction, RV32I and RV64I.

For each bit set in the five-bit zero-extended immediate, set the corresponding bit in control and status register *csr*. Expands to **csrrsi** x0, csr, zimm.

---

csrw csr, rs1 　　　　　　　　　　CSRs[csr] = x[rs1]

Control and Status Register Write. Pseudoinstruction, RV32I and RV64I.

Copies x[*rs1*] to control and status register *csr*. Expands to **csrrw** x0, csr, rs1.

---

csrwi csr, zimm[4:0] 　　　　　　　　CSRs[csr] = zimm

Control and Status Register Write Immediate. Pseudoinstruction, RV32I and RV64I.

Copies the five-bit zero-extended immediate to control and status register *csr*. Expands to **csrrwi** x0, csr, zimm.

---

div rd, rs1, rs2 　　　　　　　x[rd] = x[rs1] $\div_s$ x[rs2]

Divide. R-type, RV32M and RV64M.

Divides x[*rs1*] by x[*rs2*], rounding towards zero, treating the values as two's complement numbers, and writes the quotient to x[*rd*].

31　　　　　25	24　　　20	19　　15	14　12	11　　7	6　　　　0
0000001	rs2	rs1	100	rd	0110011

---

divu rd, rs1, rs2 　　　　　　x[rd] = x[rs1] $\div_u$ x[rs2]

Divide, Unsigned. R-type, RV32M and RV64M.

Divides x[*rs1*] by x[*rs2*], rounding towards zero, treating the values as unsigned numbers, and writes the quotient to x[*rd*].

31　　　　　25	24　　　20	19　　15	14　12	11　　7	6　　　　0
0000001	rs2	rs1	101	rd	0110011

divuw rd, rs1, rs2   x[rd] = sext(x[rs1][31:0] ÷$_u$ x[rs2][31:0])

Divide Word, Unsigned. R-type, RV64M only.

Divides the lower 32 bits of x[*rs1*] by the lower 32 bits of x[*rs2*], rounding towards zero, treating the values as unsigned numbers, and writes the sign-extended 32-bit quotient to x[*rd*].

31 25	24 20	19 15	14 12	11 7	6 0
0000001	rs2	rs1	101	rd	0111011

divw rd, rs1, rs2   x[rd] = sext(x[rs1][31:0] ÷$_s$ x[rs2][31:0])

Divide Word. R-type, RV64M only.

Divides the lower 32 bits of x[*rs1*] by the lower 32 bits of x[*rs2*], rounding towards zero, treating the values as two's complement numbers, and writes the sign-extended 32-bit quotient to x[*rd*].

31 25	24 20	19 15	14 12	11 7	6 0
0000001	rs2	rs1	100	rd	0111011

ebreak   RaiseException(Breakpoint)

Environment Breakpoint. I-type, RV32I and RV64I.

Makes a request of the debugger by raising a Breakpoint exception.

31 20	19 15	14 12	11 7	6 0
000000000001	00000	000	00000	1110011

ecall   RaiseException(EnvironmentCall)

Environment Call. I-type, RV32I and RV64I.

Makes a request of the execution environment by raising an Environment Call exception.

31 20	19 15	14 12	11 7	6 0
000000000000	00000	000	00000	1110011

fabs.d rd, rs1   f[rd] = |f[rs1]|

Floating-point Absolute Value. Pseudoinstruction, RV32D and RV64D.

Writes the absolute value of the double-precision floating-point number in f[*rs1*] to f[*rd*].
Expands to **fsgnjx.d** rd, rs1, rs1.

fabs.s rd, rs1   f[rd] = |f[rs1]|

Floating-point Absolute Value. Pseudoinstruction, RV32F and RV64F.

Writes the absolute value of the single-precision floating-point number in f[*rs1*] to f[*rd*].
Expands to **fsgnjx.s** rd, rs1, rs1.

fadd.d rd, rs1, rs2
$$f[rd] = f[rs1] + f[rs2]$$

Floating-point Add, Double-Precision. R-type, RV32D and RV64D.

Adds the double-precision floating-point numbers in registers f[*rs1*] and f[*rs2*] and writes the rounded double-precision sum to f[*rd*].

31	25 24	20 19	15 14	12 11	7 6	0
0000001	rs2	rs1	rm	rd	1010011	

---

fadd.s rd, rs1, rs2
$$f[rd] = f[rs1] + f[rs2]$$

Floating-point Add, Single-Precision. R-type, RV32F and RV64F.

Adds the single-precision floating-point numbers in registers f[*rs1*] and f[*rs2*] and writes the rounded single-precision sum to f[*rd*].

31	25 24	20 19	15 14	12 11	7 6	0
0000000	rs2	rs1	rm	rd	1010011	

---

fclass.d rd, rs1, rs2
$$x[rd] = \text{classify}_d(f[rs1])$$

Floating-point Classify, Double-Precision. R-type, RV32D and RV64D.

Writes to x[*rd*] a mask indicating the class of the double-precision floating-point number in f[*rs1*]. See the description of **fclass.s** for the interpretation of the value written to x[*rd*].

31	25 24	20 19	15 14	12 11	7 6	0
1110001	00000	rs1	001	rd	1010011	

---

fclass.s rd, rs1, rs2
$$x[rd] = \text{classify}_s(f[rs1])$$

Floating-point Classify, Single-Precision. R-type, RV32F and RV64F.

Writes to x[*rd*] a mask indicating the class of the single-precision floating-point number in f[*rs1*]. Exactly one bit in x[*rd*] is set, per the following table:

x[*rd*] bit	Meaning
0	f[*rs1*] is $-\infty$.
1	f[*rs1*] is a negative normal number.
2	f[*rs1*] is a negative subnormal number.
3	f[*rs1*] is -0.
4	f[*rs1*] is $+0$.
5	f[*rs1*] is a positive subnormal number.
6	f[*rs1*] is a positive normal number.
7	f[*rs1*] is $+\infty$.
8	f[*rs1*] is a signaling NaN.
9	f[*rs1*] is a quiet NaN.

31	25 24	20 19	15 14	12 11	7 6	0
1110000	00000	rs1	001	rd	1010011	

fcvt.d.l rd, rs1, rs2 $f[rd] = f64_{s64}(x[rs1])$

Floating-point Convert to Double from Long. R-type, RV64D only.
Converts the 64-bit two's complement integer in x[*rs1*] to a double-precision floating-point number and writes it to f[*rd*].

31	25 24	20 19	15 14	12 11	7 6	0
1101001	00010	rs1	rm	rd	1010011	

fcvt.d.lu rd, rs1, rs2 $f[rd] = f64_{u64}(x[rs1])$

Floating-point Convert to Double from Unsigned Long. R-type, RV64D only.
Converts the 64-bit unsigned integer in x[*rs1*] to a double-precision floating-point number and writes it to f[*rd*].

31	25 24	20 19	15 14	12 11	7 6	0
1101001	00011	rs1	rm	rd	1010011	

fcvt.d.s rd, rs1, rs2 $f[rd] = f64_{f32}(f[rs1])$

Floating-point Convert to Double from Single. R-type, RV32D and RV64D.
Converts the single-precision floating-point number in f[*rs1*] to a double-precision floating-point number and writes it to f[*rd*].

31	25 24	20 19	15 14	12 11	7 6	0
0100001	00000	rs1	rm	rd	1010011	

fcvt.d.w rd, rs1, rs2 $f[rd] = f64_{s32}(x[rs1])$

Floating-point Convert to Double from Word. R-type, RV32D and RV64D.
Converts the 32-bit two's complement integer in x[*rs1*] to a double-precision floating-point number and writes it to f[*rd*].

31	25 24	20 19	15 14	12 11	7 6	0
1101001	00000	rs1	rm	rd	1010011	

fcvt.d.wu rd, rs1, rs2 $f[rd] = f64_{u32}(x[rs1])$

Floating-point Convert to Double from Unsigned Word. R-type, RV32D and RV64D.
Converts the 32-bit unsigned integer in x[*rs1*] to a double-precision floating-point number and writes it to f[*rd*].

31	25 24	20 19	15 14	12 11	7 6	0
1101001	00001	rs1	rm	rd	1010011	

fcvt.l.d rd, rs1, rs2   x[rd] = s64$_{f64}$(f[rs1])

Floating-point Convert to Long from Double. R-type, RV64D only.
Converts the double-precision floating-point number in register f[*rs1*] to a 64-bit two's complement integer and writes it to x[*rd*].

31	25 24	20 19	15 14	12 11	7 6	0
1100001	00010	rs1	rm	rd	1010011	

fcvt.l.s rd, rs1, rs2   x[rd] = s64$_{f32}$(f[rs1])

Floating-point Convert to Long from Single. R-type, RV64F only.
Converts the single-precision floating-point number in register f[*rs1*] to a 64-bit two's complement integer and writes it to x[*rd*].

31	25 24	20 19	15 14	12 11	7 6	0
1100000	00010	rs1	rm	rd	1010011	

fcvt.lu.d rd, rs1, rs2   x[rd] = u64$_{f64}$(f[rs1])

Floating-point Convert to Unsigned Long from Double. R-type, RV64D only.
Converts the double-precision floating-point number in register f[*rs1*] to a 64-bit unsigned integer and writes it to x[*rd*].

31	25 24	20 19	15 14	12 11	7 6	0
1100001	00011	rs1	rm	rd	1010011	

fcvt.lu.s rd, rs1, rs2   x[rd] = u64$_{f32}$(f[rs1])

Floating-point Convert to Unsigned Long from Single. R-type, RV64F only.
Converts the single-precision floating-point number in register f[*rs1*] to a 64-bit unsigned integer and writes it to x[*rd*].

31	25 24	20 19	15 14	12 11	7 6	0
1100000	00011	rs1	rm	rd	1010011	

fcvt.s.d rd, rs1, rs2   f[rd] = f32$_{f64}$(f[rs1])

Floating-point Convert to Single from Double. R-type, RV32D and RV64D.
Converts the double-precision floating-point number in f[*rs1*] to a single-precision floating-point number and writes it to f[*rd*].

31	25 24	20 19	15 14	12 11	7 6	0
0100000	00001	rs1	rm	rd	1010011	

fcvt.s.l rd, rs1, rs2   f[rd] = f32$_{s64}$(x[rs1])

Floating-point Convert to Single from Long. R-type, RV64F only.

Converts the 64-bit two's complement integer in x[*rs1*] to a single-precision floating-point number and writes it to f[*rd*].

31	25 24	20 19	15 14	12 11	7 6	0
1101000	00010	rs1	rm	rd	1010011	

fcvt.s.lu rd, rs1, rs2   f[rd] = f32$_{u64}$(x[rs1])

Floating-point Convert to Single from Unsigned Long. R-type, RV64F only.

Converts the 64-bit unsigned integer in x[*rs1*] to a single-precision floating-point number and writes it to f[*rd*].

31	25 24	20 19	15 14	12 11	7 6	0
1101000	00011	rs1	rm	rd	1010011	

fcvt.s.w rd, rs1, rs2   f[rd] = f32$_{s32}$(x[rs1])

Floating-point Convert to Single from Word. R-type, RV32F and RV64F.

Converts the 32-bit two's complement integer in x[*rs1*] to a single-precision floating-point number and writes it to f[*rd*].

31	25 24	20 19	15 14	12 11	7 6	0
1101000	00000	rs1	rm	rd	1010011	

fcvt.s.wu rd, rs1, rs2   f[rd] = f32$_{u32}$(x[rs1])

Floating-point Convert to Single from Unsigned Word. R-type, RV32F and RV64F.

Converts the 32-bit unsigned integer in x[*rs1*] to a single-precision floating-point number and writes it to f[*rd*].

31	25 24	20 19	15 14	12 11	7 6	0
1101000	00001	rs1	rm	rd	1010011	

fcvt.w.d rd, rs1, rs2   x[rd] = sext(s32$_{f64}$(f[rs1]))

Floating-point Convert to Word from Double. R-type, RV32D and RV64D.

Converts the double-precision floating-point number in register f[*rs1*] to a 32-bit two's complement integer and writes the sign-extended result to x[*rd*].

31	25 24	20 19	15 14	12 11	7 6	0
1100001	00000	rs1	rm	rd	1010011	

fcvt.wu.d rd, rs1, rs2 $\qquad$ x[rd] = sext(u32$_{f64}$(f[rs1]))

Floating-point Convert to Unsigned Word from Double. R-type, RV32D and RV64D.
Converts the double-precision floating-point number in register f[*rs1*] to a 32-bit unsigned
integer and writes the sign-extended result to x[*rd*].

31	25 24	20 19	15 14	12 11	7 6	0
1100001	00001	rs1	rm	rd	1010011	

fcvt.w.s rd, rs1, rs2 $\qquad$ x[rd] = sext(s32$_{f32}$(f[rs1]))

Floating-point Convert to Word from Single. R-type, RV32F and RV64F.
Converts the single-precision floating-point number in register f[*rs1*] to a 32-bit two's complement integer and writes the sign-extended result to x[*rd*].

31	25 24	20 19	15 14	12 11	7 6	0
1100000	00000	rs1	rm	rd	1010011	

fcvt.wu.s rd, rs1, rs2 $\qquad$ x[rd] = sext(u32$_{f32}$(f[rs1]))

Floating-point Convert to Unsigned Word from Single. R-type, RV32F and RV64F.
Converts the single-precision floating-point number in register f[*rs1*] to a 32-bit unsigned
integer and writes the sign-extended result to x[*rd*].

31	25 24	20 19	15 14	12 11	7 6	0
1100000	00001	rs1	rm	rd	1010011	

fdiv.d rd, rs1, rs2 $\qquad$ f[rd] = f[rs1] ÷ f[rs2]

Floating-point Divide, Double-Precision. R-type, RV32D and RV64D.
Divides the double-precision floating-point number in register f[*rs1*] by f[*rs2*] and writes the
rounded double-precision quotient to f[*rd*].

31	25 24	20 19	15 14	12 11	7 6	0
0001101	rs2	rs1	rm	rd	1010011	

fdiv.s rd, rs1, rs2 $\qquad$ f[rd] = f[rs1] ÷ f[rs2]

Floating-point Divide, Single-Precision. R-type, RV32F and RV64F.
Divides the single-precision floating-point number in register f[*rs1*] by f[*rs2*] and writes the
rounded single-precision quotient to f[*rd*].

31	25 24	20 19	15 14	12 11	7 6	0
0001100	rs2	rs1	rm	rd	1010011	

fence pred, succ   Fence(pred, succ)

Fence Memory and I/O. I-type, RV32I and RV64I.

Renders preceding memory and I/O accesses in the *pred*ecessor set observable to other threads and devices before subsequent memory and I/O accesses in the *succ*essor set become observable. Bits 3, 2, 1, and 0 in these sets correspond to device **i**nput, device **o**utput, memory **r**eads, and memory **w**rites, respectively. The instruction **fence** r,rw, for example, orders older reads with younger reads and writes, and is encoded with *pred*=0010 and *succ*=0011. If the arguments are omitted, a full **fence** iorw, iorw is implied.

31 28	27 24	23 20	19 15	14 12	11 7	6 0
0000	pred	succ	00000	000	00000	0001111

fence.i   Fence(Store, Fetch)

Fence Instruction Stream. I-type, RV32I and RV64I.

Renders stores to instruction memory observable to subsequent instruction fetches.

31	20	19 15	14 12	11 7	6 0
000000000000		00000	001	00000	0001111

feq.d rd, rs1, rs2   x[rd] = f[rs1] == f[rs2]

Floating-point Equals, Double-Precision. R-type, RV32D and RV64D.

Writes 1 to x[*rd*] if the double-precision floating-point number in f[*rs1*] equals the number in f[*rs2*], and 0 if not.

31 25	24 20	19 15	14 12	11 7	6 0
1010001	rs2	rs1	010	rd	1010011

feq.s rd, rs1, rs2   x[rd] = f[rs1] == f[rs2]

Floating-point Equals, Single-Precision. R-type, RV32F and RV64F.

Writes 1 to x[*rd*] if the single-precision floating-point number in f[*rs1*] equals the number in f[*rs2*], and 0 if not.

31 25	24 20	19 15	14 12	11 7	6 0
1010000	rs2	rs1	010	rd	1010011

fld rd, offset(rs1)   f[rd] = M[x[rs1] + sext(offset)][63:0]

Floating-point Load Doubleword. I-type, RV32D and RV64D.

Loads a double-precision floating-point number from memory address x[*rs1*] + *sign-extend*(*offset*) and writes it to f[*rd*].

Compressed forms: **c.fldsp** rd, offset; **c.fld** rd, offset(rs1)

31	20	19 15	14 12	11 7	6 0
offset[11:0]		rs1	011	rd	0000111

fle.d rd, rs1, rs2   x[rd] = f[rs1] $\leq$ f[rs2]

Floating-point Less Than or Equal, Double-Precision. R-type, RV32D and RV64D.

Writes 1 to x[*rd*] if the double-precision floating-point number in f[*rs1*] is less than or equal to the number in f[*rs2*], and 0 if not.

31	25 24	20 19	15 14	12 11	7 6	0
1010001	rs2	rs1	000	rd	1010011	

fle.s rd, rs1, rs2   x[rd] = f[rs1] $\leq$ f[rs2]

Floating-point Less Than or Equal, Single-Precision. R-type, RV32F and RV64F.

Writes 1 to x[*rd*] if the single-precision floating-point number in f[*rs1*] is less than or equal to the number in f[*rs2*], and 0 if not.

31	25 24	20 19	15 14	12 11	7 6	0
1010000	rs2	rs1	000	rd	1010011	

flt.d rd, rs1, rs2   x[rd] = f[rs1] < f[rs2]

Floating-point Less Than, Double-Precision. R-type, RV32D and RV64D.

Writes 1 to x[*rd*] if the double-precision floating-point number in f[*rs1*] is less than the number in f[*rs2*], and 0 if not.

31	25 24	20 19	15 14	12 11	7 6	0
1010001	rs2	rs1	001	rd	1010011	

flt.s rd, rs1, rs2   x[rd] = f[rs1] < f[rs2]

Floating-point Less Than, Single-Precision. R-type, RV32F and RV64F.

Writes 1 to x[*rd*] if the single-precision floating-point number in f[*rs1*] is less than the number in f[*rs2*], and 0 if not.

31	25 24	20 19	15 14	12 11	7 6	0
1010000	rs2	rs1	001	rd	1010011	

flw rd, offset(rs1)   f[rd] = M[x[rs1] + sext(offset)][31:0]

Floating-point Load Word. I-type, RV32F and RV64F.

Loads a single-precision floating-point number from memory address x[*rs1*] + *sign-extend*(*offset*) and writes it to f[*rd*].

Compressed forms: **c.flwsp** rd, offset; **c.flw** rd, offset(rs1)

31	20 19	15 14	12 11	7 6	0
offset[11:0]	rs1	010	rd	0000111	

fmadd.d rd, rs1, rs2, rs3   f[rd] = f[rs1]×f[rs2]+f[rs3]

Floating-point Fused Multiply-Add, Double-Precision. R4-type, RV32D and RV64D.
Multiplies the double-precision floating-point numbers in f[*rs1*] and f[*rs2*], adds the unrounded product to the double-precision floating-point number in f[*rs3*], and writes the
rounded double-precision result to f[*rd*].

31	27 26	25 24	20 19	15 14	12 11	7 6	0
rs3	01	rs2	rs1	rm	rd	1000011	

fmadd.s rd, rs1, rs2, rs3   f[rd] = f[rs1]×f[rs2]+f[rs3]

Floating-point Fused Multiply-Add, Single-Precision. R4-type, RV32F and RV64F.
Multiplies the single-precision floating-point numbers in f[*rs1*] and f[*rs2*], adds the unrounded product to the single-precision floating-point number in f[*rs3*], and writes the
rounded single-precision result to f[*rd*].

31	27 26	25 24	20 19	15 14	12 11	7 6	0
rs3	00	rs2	rs1	rm	rd	1000011	

fmax.d rd, rs1, rs2   f[rd] = max(f[rs1], f[rs2])

Floating-point Maximum, Double-Precision. R-type, RV32D and RV64D.
Copies the larger of the double-precision floating-point numbers in registers f[*rs1*] and f[*rs2*]
to f[*rd*].

31	25 24	20 19	15 14	12 11	7 6	0
0010101	rs2	rs1	001	rd	1010011	

fmax.s rd, rs1, rs2   f[rd] = max(f[rs1], f[rs2])

Floating-point Maximum, Single-Precision. R-type, RV32F and RV64F.
Copies the larger of the single-precision floating-point numbers in registers f[*rs1*] and f[*rs2*]
to f[*rd*].

31	25 24	20 19	15 14	12 11	7 6	0
0010100	rs2	rs1	001	rd	1010011	

fmin.d rd, rs1, rs2   f[rd] = min(f[rs1], f[rs2])

Floating-point Minimum, Double-Precision. R-type, RV32D and RV64D.
Copies the smaller of the double-precision floating-point numbers in registers f[*rs1*] and
f[*rs2*] to f[*rd*].

31	25 24	20 19	15 14	12 11	7 6	0
0010101	rs2	rs1	000	rd	1010011	

fmin.s rd, rs1, rs2   f[rd] = min(f[rs1], f[rs2])

Floating-point Minimum, Single-Precision. R-type, RV32F and RV64F.
Copies the smaller of the single-precision floating-point numbers in registers f[*rs1*] and f[*rs2*] to f[*rd*].

31 25	24 20	19 15	14 12	11 7	6 0
0010100	rs2	rs1	000	rd	1010011

fmsub.d rd, rs1, rs2, rs3   f[rd] = f[rs1]×f[rs2]-f[rs3]

Floating-point Fused Multiply-Subtract, Double-Precision. R4-type, RV32D and RV64D.
Multiplies the double-precision floating-point numbers in f[*rs1*] and f[*rs2*], subtracts the double-precision floating-point number in f[*rs3*] from the unrounded product, and writes the rounded double-precision result to f[*rd*].

31 27	26 25	24 20	19 15	14 12	11 7	6 0
rs3	01	rs2	rs1	rm	rd	1000111

fmsub.s rd, rs1, rs2, rs3   f[rd] = f[rs1]×f[rs2]-f[rs3]

Floating-point Fused Multiply-Subtract, Single-Precision. R4-type, RV32F and RV64F.
Multiplies the single-precision floating-point numbers in f[*rs1*] and f[*rs2*], subtracts the single-precision floating-point number in f[*rs3*] from the unrounded product, and writes the rounded single-precision result to f[*rd*].

31 27	26 25	24 20	19 15	14 12	11 7	6 0
rs3	00	rs2	rs1	rm	rd	1000111

fmul.d rd, rs1, rs2   f[rd] = f[rs1] × f[rs2]

Floating-point Multiply, Double-Precision. R-type, RV32D and RV64D.
Multiplies the double-precision floating-point numbers in registers f[*rs1*] and f[*rs2*] and writes the rounded double-precision product to f[*rd*].

31 25	24 20	19 15	14 12	11 7	6 0
0001001	rs2	rs1	rm	rd	1010011

fmul.s rd, rs1, rs2   f[rd] = f[rs1] × f[rs2]

Floating-point Multiply, Single-Precision. R-type, RV32F and RV64F.
Multiplies the single-precision floating-point numbers in registers f[*rs1*] and f[*rs2*] and writes the rounded single-precision product to f[*rd*].

31 25	24 20	19 15	14 12	11 7	6 0
0001000	rs2	rs1	rm	rd	1010011

fmv.d rd, rs1   `f[rd] = f[rs1]`

Floating-point Move. Pseudoinstruction, RV32D and RV64D.
Copies the double-precision floating-point number in f[*rs1*] to f[*rd*]. Expands to **fsgnj.d** rd, rs1, rs1.

fmv.d.x rd, rs1, rs2   `f[rd] = x[rs1][63:0]`

Floating-point Move Doubleword from Integer. R-type, RV64D only.
Copies the double-precision floating-point number in register x[*rs1*] to f[*rd*].

31	25 24	20 19	15 14	12 11	7 6	0
1111001	00000	rs1	000	rd	1010011	

fmv.s rd, rs1   `f[rd] = f[rs1]`

Floating-point Move. Pseudoinstruction, RV32F and RV64F.
Copies the single-precision floating-point number in f[*rs1*] to f[*rd*]. Expands to **fsgnj.s** rd, rs1, rs1.

fmv.w.x rd, rs1, rs2   `f[rd] = x[rs1][31:0]`

Floating-point Move Word from Integer. R-type, RV32F and RV64F.
Copies the single-precision floating-point number in register x[*rs1*] to f[*rd*].

31	25 24	20 19	15 14	12 11	7 6	0
1111000	00000	rs1	000	rd	1010011	

fmv.x.d rd, rs1, rs2   `x[rd] = f[rs1][63:0]`

Floating-point Move Doubleword to Integer. R-type, RV64D only.
Copies the double-precision floating-point number in register f[*rs1*] to x[*rd*].

31	25 24	20 19	15 14	12 11	7 6	0
1110001	00000	rs1	000	rd	1010011	

fmv.x.w rd, rs1, rs2   `x[rd] = sext(f[rs1][31:0])`

Floating-point Move Word to Integer. R-type, RV32F and RV64F.
Copies the single-precision floating-point number in register f[*rs1*] to x[*rd*], sign-extending the result for RV64F.

31	25 24	20 19	15 14	12 11	7 6	0
1110000	00000	rs1	000	rd	1010011	

fneg.d rd, rs1   `f[rd] = -f[rs1]`

Floating-point Negate. Pseudoinstruction, RV32D and RV64D.

Writes the opposite of the double-precision floating-point number in f[*rs1*] to f[*rd*]. Expands to **fsgnjn.d** rd, rs1, rs1.

---

fneg.s rd, rs1   `f[rd] = -f[rs1]`

Floating-point Negate. Pseudoinstruction, RV32F and RV64F.

Writes the opposite of the single-precision floating-point number in f[*rs1*] to f[*rd*]. Expands to **fsgnjn.s** rd, rs1, rs1.

---

fnmadd.d rd, rs1, rs2, rs3   `f[rd] = -f[rs1]×f[rs2]-f[rs3]`

Floating-point Fused Negative Multiply-Add, Double-Precision. R4-type, RV32D and RV64D.

Multiplies the double-precision floating-point numbers in f[*rs1*] and f[*rs2*], negates the result, subtracts the double-precision floating-point number in f[*rs3*] from the unrounded product, and writes the rounded double-precision result to f[*rd*].

31	27 26 25 24	20 19	15 14 12 11	7 6	0
rs3	01	rs2	rs1	rm	rd	1001111

---

fnmadd.s rd, rs1, rs2, rs3   `f[rd] = -f[rs1]×f[rs2]-f[rs3]`

Floating-point Fused Negative Multiply-Add, Single-Precision. R4-type, RV32F and RV64F.

Multiplies the single-precision floating-point numbers in f[*rs1*] and f[*rs2*], negates the result, subtracts the single-precision floating-point number in f[*rs3*] from the unrounded product, and writes the rounded single-precision result to f[*rd*].

31	27 26 25 24	20 19	15 14 12 11	7 6	0
rs3	00	rs2	rs1	rm	rd	1001111

---

fnmsub.d rd, rs1, rs2, rs3   `f[rd] = -f[rs1]×f[rs2]+f[rs3]`

Floating-point Fused Negative Multiply-Subtract, Double-Precision. R4-type, RV32D and RV64D.

Multiplies the double-precision floating-point numbers in f[*rs1*] and f[*rs2*], negates the result, adds the unrounded product to the double-precision floating-point number in f[*rs3*], and writes the rounded double-precision result to f[*rd*].

31	27 26 25 24	20 19	15 14 12 11	7 6	0
rs3	01	rs2	rs1	rm	rd	1001011

fnmsub.s rd, rs1, rs2, rs3   f[rd] = -f[rs1]×f[rs2]+f[rs3]

Floating-point Fused Negative Multiply-Subtract, Single-Precision. R4-type, RV32F and RV64F.

Multiplies the single-precision floating-point numbers in f[*rs1*] and f[*rs2*], negates the result, adds the unrounded product to the single-precision floating-point number in f[*rs3*], and writes the rounded single-precision result to f[*rd*].

31	27 26 25 24	20 19	15 14	12 11	7 6	0
rs3	00	rs2	rs1	rm	rd	1001011

frcsr rd   x[rd] = CSRs[fcsr]

Floating-point Read Control and Status Register. Pseudoinstruction, RV32F and RV64F.

Copies the floating-point control and status register to x[*rd*]. Expands to **csrrs** rd, fcsr, x0.

frflags rd   x[rd] = CSRs[fflags]

Floating-point Read Exception Flags. Pseudoinstruction, RV32F and RV64F.

Copies the floating-point exception flags to x[*rd*]. Expands to **csrrs** rd, fflags, x0.

frrm rd   x[rd] = CSRs[frm]

Floating-point Read Rounding Mode. Pseudoinstruction, RV32F and RV64F.

Copies the floating-point rounding mode to x[*rd*]. Expands to **csrrs** rd, frm, x0.

fscsr rd, rs1   t = CSRs[fcsr]; CSRs[fcsr] = x[rs1]; x[rd] = t

Floating-point Swap Control and Status Register. Pseudoinstruction, RV32F and RV64F.

Copies x[*rs1*] to the floating-point control and status register, then copies the previous value of the floating-point control and status register to x[*rd*]. Expands to **csrrw** rd, fcsr, rs1. If *rd* is omitted, x0 is assumed.

fsd rs2, offset(rs1)   M[x[rs1] + sext(offset)] = f[rs2][63:0]

Floating-point Store Doubleword. S-type, RV32D and RV64D.

Stores the double-precision floating-point number in register f[*rs2*] to memory at address x[*rs1*] + *sign-extend*(*offset*).

Compressed forms: **c.fsdsp** rs2, offset; **c.fsd** rs2, offset(rs1)

31	25 24	20 19	15 14	12 11	7 6	0
offset[11:5]	rs2	rs1	011	offset[4:0]	0100111	

fsflags rd, rs1   `t = CSRs[fflags]; CSRs[fflags] = x[rs1]; x[rd] = t`

Floating-point Swap Exception Flags. Pseudoinstruction, RV32F and RV64F.

Copies x[*rs1*] to the floating-point exception flags register, then copies the previous floating-point exception flags to x[*rd*]. Expands to **csrrw** rd, fflags, rs1. If *rd* is omitted, x0 is assumed.

---

fsgnj.d rd, rs1, rs2   `f[rd] = {f[rs2][63], f[rs1][62:0]}`

Floating-point Sign Inject, Double-Precision. R-type, RV32D and RV64D.

Constructs a new double-precision floating-point number from the exponent and significand of f[*rs1*], taking the sign from f[*rs2*], and writes it to f[*rd*].

31 25	24 20	19 15	14 12	11 7	6 0
0010001	rs2	rs1	000	rd	1010011

---

fsgnj.s rd, rs1, rs2   `f[rd] = {f[rs2][31], f[rs1][30:0]}`

Floating-point Sign Inject, Single-Precision. R-type, RV32F and RV64F.

Constructs a new single-precision floating-point number from the exponent and significand of f[*rs1*], taking the sign from f[*rs2*], and writes it to f[*rd*].

31 25	24 20	19 15	14 12	11 7	6 0
0010000	rs2	rs1	000	rd	1010011

---

fsgnjn.d rd, rs1, rs2   `f[rd] = {~f[rs2][63], f[rs1][62:0]}`

Floating-point Sign Inject-Negate, Double-Precision. R-type, RV32D and RV64D.

Constructs a new double-precision floating-point number from the exponent and significand of f[*rs1*], taking the opposite sign of f[*rs2*], and writes it to f[*rd*].

31 25	24 20	19 15	14 12	11 7	6 0
0010001	rs2	rs1	001	rd	1010011

---

fsgnjn.s rd, rs1, rs2   `f[rd] = {~f[rs2][31], f[rs1][30:0]}`

Floating-point Sign Inject-Negate, Single-Precision. R-type, RV32F and RV64F.

Constructs a new single-precision floating-point number from the exponent and significand of f[*rs1*], taking the opposite sign of f[*rs2*], and writes it to f[*rd*].

31 25	24 20	19 15	14 12	11 7	6 0
0010000	rs2	rs1	001	rd	1010011

---

fsgnjx.d rd, rs1, rs2 f[rd] = {f[rs1][63] ^ f[rs2][63], f[rs1][62:0]}

Floating-point Sign Inject-XOR, Double-Precision. R-type, RV32D and RV64D.

Constructs a new double-precision floating-point number from the exponent and significand of f[*rs1*], taking the sign from the XOR of the signs of f[*rs1*] and f[*rs2*], and writes it to f[*rd*].

31	25 24	20 19	15 14	12 11	7 6	0
0010001	rs2	rs1	010	rd	1010011	

fsgnjx.s rd, rs1, rs2 f[rd] = {f[rs1][31] ^ f[rs2][31], f[rs1][30:0]}

Floating-point Sign Inject-XOR, Single-Precision. R-type, RV32F and RV64F.

Constructs a new single-precision floating-point number from the exponent and significand of f[*rs1*], taking the sign from the XOR of the signs of f[*rs1*] and f[*rs2*], and writes it to f[*rd*].

31	25 24	20 19	15 14	12 11	7 6	0
0010000	rs2	rs1	010	rd	1010011	

fsqrt.d rd, rs1, rs2   f[rd] = $\sqrt{\text{f[rs1]}}$

Floating-point Square Root, Double-Precision. R-type, RV32D and RV64D.

Computes the square root of the double-precision floating-point number in register f[*rs1*] and writes the rounded double-precision result to f[*rd*].

31	25 24	20 19	15 14	12 11	7 6	0
0101101	00000	rs1	rm	rd	1010011	

fsqrt.s rd, rs1, rs2   f[rd] = $\sqrt{\text{f[rs1]}}$

Floating-point Square Root, Single-Precision. R-type, RV32F and RV64F.

Computes the square root of the single-precision floating-point number in register f[*rs1*] and writes the rounded single-precision result to f[*rd*].

31	25 24	20 19	15 14	12 11	7 6	0
0101100	00000	rs1	rm	rd	1010011	

fsrm rd, rs1   t = CSRs[frm]; CSRs[frm] = x[rs1]; x[rd] = t

Floating-point Swap Rounding Mode. Pseudoinstruction, RV32F and RV64F.

Copies x[*rs1*] to the floating-point rounding mode register, then copies the previous floating-point rounding mode to x[*rd*]. Expands to **csrrw** rd, frm, rs1. If *rd* is omitted, x0 is assumed.

fsub.d rd, rs1, rs2   f[rd] = f[rs1] - f[rs2]

Floating-point Subtract, Double-Precision. R-type, RV32D and RV64D.
Subtracts the double-precision floating-point number in register f[*rs2*] from f[*rs1*] and writes
the rounded double-precision difference to f[*rd*].

31	25 24	20 19	15 14	12 11	7 6	0
0000101	rs2	rs1	rm	rd	1010011	

fsub.s rd, rs1, rs2   f[rd] = f[rs1] - f[rs2]

Floating-point Subtract, Single-Precision. R-type, RV32F and RV64F.
Subtracts the single-precision floating-point number in register f[*rs2*] from f[*rs1*] and writes
the rounded single-precision difference to f[*rd*].

31	25 24	20 19	15 14	12 11	7 6	0
0000100	rs2	rs1	rm	rd	1010011	

fsw rs2, offset(rs1)   M[x[rs1] + sext(offset)] = f[rs2][31:0]

Floating-point Store Word. S-type, RV32F and RV64F.
Stores the single-precision floating-point number in register f[*rs2*] to memory at address
x[*rs1*] + *sign-extend*(*offset*).
Compressed forms: **c.fswsp** rs2, offset; **c.fsw** rs2, offset(rs1)

31	25 24	20 19	15 14	12 11	7 6	0
offset[11:5]	rs2	rs1	010	offset[4:0]	0100111	

j offset   pc += sext(offset)

Jump. Pseudoinstruction, RV32I and RV64I.
Sets the *pc* to the current *pc* plus the sign-extended *offset*. Expands to **jal** x0, offset.

jal rd, offset   x[rd] = pc+4; pc += sext(offset)

Jump and Link. J-type, RV32I and RV64I.
Writes the address of the next instruction (*pc*+4) to x[*rd*], then set the *pc* to the current *pc*
plus the sign-extended *offset*. If *rd* is omitted, x1 is assumed.
Compressed forms: **c.j** offset; **c.jal** offset

31	12 11	7 6	0
offset[20|10:1|11|19:12]	rd	1101111	

jalr rd, offset(rs1) t=pc+4; pc=(x[rs1]+sext(offset))&~1; x[rd]=t

Jump and Link Register. I-type, RV32I and RV64I.

Sets the *pc* to x[*rs1*] + *sign-extend*(*offset*), masking off the least-significant bit of the computed address, then writes the previous *pc*+4 to x[*rd*]. If *rd* is omitted, x1 is assumed.

Compressed forms: **c.jr** rs1; **c.jalr** rs1

31 20	19 15	14 12	11 7	6 0
offset[11:0]	rs1	000	rd	1100111

jr rs1   pc = x[rs1]

Jump Register. Pseudoinstruction, RV32I and RV64I.

Sets the *pc* to x[*rs1*]. Expands to **jalr** x0, 0(rs1).

la rd, symbol   x[rd] = &symbol

Load Address. Pseudoinstruction, RV32I and RV64I.

Loads the address of *symbol* into x[*rd*]. When assembling position-independent code, it expands into a load from the Global Offset Table: for RV32I, **auipc** rd, offsetHi then **lw** rd, offsetLo(rd); for RV64I, **auipc** rd, offsetHi then **ld** rd, offsetLo(rd). Otherwise, it expands into **auipc** rd, offsetHi then **addi** rd, rd, offsetLo.

lb rd, offset(rs1)   x[rd] = sext(M[x[rs1] + sext(offset)][7:0])

Load Byte. I-type, RV32I and RV64I.

Loads a byte from memory at address x[*rs1*] + *sign-extend*(*offset*) and writes it to x[*rd*], sign-extending the result.

31 20	19 15	14 12	11 7	6 0
offset[11:0]	rs1	000	rd	0000011

lbu rd, offset(rs1)   x[rd] = M[x[rs1] + sext(offset)][7:0]

Load Byte, Unsigned. I-type, RV32I and RV64I.

Loads a byte from memory at address x[*rs1*] + *sign-extend*(*offset*) and writes it to x[*rd*], zero-extending the result.

31 20	19 15	14 12	11 7	6 0
offset[11:0]	rs1	100	rd	0000011

ld rd, offset(rs1)　　　　　　　　x[rd] = M[x[rs1] + sext(offset)][63:0]

Load Doubleword. I-type, RV64I only.

Loads eight bytes from memory at address x[*rs1*] + *sign-extend*(*offset*) and writes them to x[*rd*].

Compressed forms: **c.ldsp** rd, offset; **c.ld** rd, offset(rs1)

offset[11:0]	rs1	011	rd	0000011

31　　　　　　　　　　20 19　　　15 14　　12 11　　　7 6　　　　　　0

---

lh rd, offset(rs1)　　　　　x[rd] = sext(M[x[rs1] + sext(offset)][15:0])

Load Halfword. I-type, RV32I and RV64I.

Loads two bytes from memory at address x[*rs1*] + *sign-extend*(*offset*) and writes them to x[*rd*], sign-extending the result.

offset[11:0]	rs1	001	rd	0000011

31　　　　　　　　　　20 19　　　15 14　　12 11　　　7 6　　　　　　0

---

lhu rd, offset(rs1)　　　　　　x[rd] = M[x[rs1] + sext(offset)][15:0]

Load Halfword, Unsigned. I-type, RV32I and RV64I.

Loads two bytes from memory at address x[*rs1*] + *sign-extend*(*offset*) and writes them to x[*rd*], zero-extending the result.

offset[11:0]	rs1	101	rd	0000011

31　　　　　　　　　　20 19　　　15 14　　12 11　　　7 6　　　　　　0

---

li rd, immediate　　　　　　　　　　　　　　　x[rd] = immediate

Load Immediate. Pseudoinstruction, RV32I and RV64I.

Loads a constant into x[*rd*], using as few instructions as possible. For RV32I, it expands to **lui** and/or **addi**; for RV64I, it's as long as **lui, addi, slli, addi, slli, addi, slli, addi**.

---

la rd, symbol　　　　　　　　　　　　　　　　x[rd] = &symbol

Load Local Address. Pseudoinstruction, RV32I and RV64I.

Loads the address of *symbol* into x[*rd*]. Expands into **auipc** rd, offsetHi then **addi** rd, rd, offsetLo.

---

lr.d rd, (rs1)　　　　　　　　x[rd] = LoadReserved64(M[x[rs1]])

Load-Reserved Doubleword. R-type, RV64A only.

Loads the eight bytes from memory at address x[*rs1*], writes them to x[*rd*], and registers a reservation on that memory doubleword.

00010	aq	rl	00000	rs1	011	rd	0101111

31　　　27 26　25 24　　　20 19　　　15 14　　12 11　　　7 6　　　　　0

lr.w rd, (rs1)   x[rd] = LoadReserved32(M[x[rs1]])

Load-Reserved Word. R-type, RV32A and RV64A.

Loads the four bytes from memory at address x[*rs1*], writes them to x[*rd*], sign-extending the result, and registers a reservation on that memory word.

31	27	26	25	24	20	19	15	14	12	11	7	6	0
00010		aq	rl	00000		rs1		010		rd		0101111	

lw rd, offset(rs1)   x[rd] = sext(M[x[rs1] + sext(offset)][31:0])

Load Word. I-type, RV32I and RV64I.

Loads four bytes from memory at address x[*rs1*] + *sign-extend*(*offset*) and writes them to x[*rd*]. For RV64I, the result is sign-extended.

Compressed forms: **c.lwsp** rd, offset; **c.lw** rd, offset(rs1)

31	20	19	15	14	12	11	7	6	0
offset[11:0]		rs1		010		rd		0000011	

lwu rd, offset(rs1)   x[rd] = M[x[rs1] + sext(offset)][31:0]

Load Word, Unsigned. I-type, RV64I only.

Loads four bytes from memory at address x[*rs1*] + *sign-extend*(*offset*) and writes them to x[*rd*], zero-extending the result.

31	20	19	15	14	12	11	7	6	0
offset[11:0]		rs1		110		rd		0000011	

lui rd, immediate   x[rd] = sext(immediate[31:12] << 12)

Load Upper Immediate. U-type, RV32I and RV64I.

Writes the sign-extended 20-bit *immediate*, left-shifted by 12 bits, to x[*rd*], zeroing the lower 12 bits.

Compressed form: **c.lui** rd, imm

31	12	11	7	6	0
immediate[31:12]		rd		0110111	

mret   ExceptionReturn(Machine)

Machine-mode Exception Return. R-type, RV32I and RV64I privileged architectures.

Returns from a machine-mode exception handler. Sets the *pc* to CSRs[mepc], the privilege mode to CSRs[mstatus].MPP, CSRs[mstatus].MIE to CSRs[mstatus].MPIE, and CSRs[mstatus].MPIE to 1; and, if user mode is supported, sets CSRs[mstatus].MPP to 0.

31	25	24	20	19	15	14	12	11	7	6	0
0011000		00010		00000		000		00000		1110011	

mul rd, rs1, rs2 $\qquad$ x[rd] = x[rs1] × x[rs2]

Multiply. R-type, RV32M and RV64M.

Multiplies x[*rs1*] by x[*rs2*] and writes the product to x[*rd*]. Arithmetic overflow is ignored.

31	25 24	20 19	15 14	12 11	7 6	0
0000001	rs2	rs1	000	rd	0110011	

mulh rd, rs1, rs2 $\qquad$ x[rd] = (x[rs1] $_s\times_s$ x[rs2]) $>>_s$ XLEN

Multiply High. R-type, RV32M and RV64M.

Multiplies x[*rs1*] by x[*rs2*], treating the values as two's complement numbers, and writes the upper half of the product to x[*rd*].

31	25 24	20 19	15 14	12 11	7 6	0
0000001	rs2	rs1	001	rd	0110011	

mulhsu rd, rs1, rs2 $\qquad$ x[rd] = (x[rs1] $_s\times_u$ x[rs2]) $>>_s$ XLEN

Multiply High Signed-Unsigned. R-type, RV32M and RV64M.

Multiplies x[*rs1*] by x[*rs2*], treating x[rs1] as a two's complement number and x[rs2] as an unsigned number, and writes the upper half of the product to x[*rd*].

31	25 24	20 19	15 14	12 11	7 6	0
0000001	rs2	rs1	010	rd	0110011	

mulhu rd, rs1, rs2 $\qquad$ x[rd] = (x[rs1] $_u\times_u$ x[rs2]) $>>_u$ XLEN

Multiply High Unsigned. R-type, RV32M and RV64M.

Multiplies x[*rs1*] by x[*rs2*], treating the values as unsigned numbers, and writes the upper half of the product to x[*rd*].

31	25 24	20 19	15 14	12 11	7 6	0
0000001	rs2	rs1	011	rd	0110011	

mulw rd, rs1, rs2 $\qquad$ x[rd] = sext((x[rs1] × x[rs2])[31:0])

Multiply Word. R-type, RV64M only.

Multiplies x[*rs1*] by x[*rs2*], truncates the product to 32 bits, and writes the sign-extended result to x[*rd*]. Arithmetic overflow is ignored.

31	25 24	20 19	15 14	12 11	7 6	0
0000001	rs2	rs1	000	rd	0111011	

mv rd, rs1 $\qquad$ x[rd] = x[rs1]

Move. Pseudoinstruction, RV32I and RV64I.

Copies register x[*rs1*] to x[*rd*]. Expands to **addi** rd, rs1, 0.

neg rd, rs2   x[rd] = -x[rs2]
Negate. Pseudoinstruction, RV32I and RV64I.
Writes the two's complement of x[*rs2*] to x[*rd*]. Expands to **sub** rd, x0, rs2.

---

negw rd, rs2   x[rd] = sext((-x[rs2])[31:0])
Negate Word. Pseudoinstruction, RV64I only.
Computes the two's complement of x[*rs2*], truncates the result to 32 bits, and writes the
sign-extended result to x[*rd*]. Expands to **subw** rd, x0, rs2.

---

nop   *Nothing*
No operation. Pseudoinstruction, RV32I and RV64I.
Merely advances the *pc* to the next instruction. Expands to **addi** x0, x0, 0.

---

not rd, rs1   x[rd] = ~x[rs1]
NOT. Pseudoinstruction, RV32I and RV64I.
Writes the ones' complement of x[*rs1*] to x[*rd*]. Expands to **xori** rd, rs1, -1.

---

or rd, rs1, rs2   x[rd] = x[rs1] | x[rs2]
OR. R-type, RV32I and RV64I.
Computes the bitwise inclusive-OR of registers x[*rs1*] and x[*rs2*] and writes the result to
x[*rd*].
Compressed form: **c.or** rd, rs2

31 25	24 20	19 15	14 12	11 7	6 0
0000000	rs2	rs1	110	rd	0110011

---

ori rd, rs1, immediate   x[rd] = x[rs1] | sext(immediate)
OR Immediate. I-type, RV32I and RV64I.
Computes the bitwise inclusive-OR of the sign-extended *immediate* and register x[*rs1*] and
writes the result to x[*rd*].

31 20	19 15	14 12	11 7	6 0
immediate[11:0]	rs1	110	rd	0010011

---

rdcycle rd   x[rd] = CSRs[cycle]
Read Cycle Counter. Pseudoinstruction, RV32I and RV64I.
Writes the number of cycles that have elapsed to x[*rd*]. Expands to **csrrs** rd, cycle, x0.

rdcycleh rd   `x[rd] = CSRs[cycleh]`

Read Cycle Counter High. Pseudoinstruction, RV32I only.
Writes the number of cycles that have elapsed, shifted right by 32 bits, to x[*rd*]. Expands to **csrrs** rd, cycleh, x0.

---

rdinstret rd   `x[rd] = CSRs[instret]`

Read Instructions-Retired Counter. Pseudoinstruction, RV32I and RV64I.
Writes the number of instructions that have retired to x[*rd*]. Expands to **csrrs** rd, instret, x0.

---

rdinstreth rd   `x[rd] = CSRs[instreth]`

Read Instructions-Retired Counter High. Pseudoinstruction, RV32I only.
Writes the number of instructions that have retired, shifted right by 32 bits, to x[*rd*]. Expands to **csrrs** rd, instreth, x0.

---

rdtime rd   `x[rd] = CSRs[time]`

Read Time. Pseudoinstruction, RV32I and RV64I.
Writes the current time to x[*rd*]. The timer frequency is platform-dependent. Expands to **csrrs** rd, time, x0.

---

rdtimeh rd   `x[rd] = CSRs[timeh]`

Read Time High. Pseudoinstruction, RV32I only.
Writes the current time, shifted right by 32 bits, to x[*rd*]. The timer frequency is platform-dependent. Expands to **csrrs** rd, timeh, x0.

---

rem rd, rs1, rs2 $x[rd] = x[rs1] \%_s x[rs2]$

Remainder. R-type, RV32M and RV64M.
Divides x[*rs1*] by x[*rs2*], rounding towards zero, treating the values as two's complement numbers, and writes the remainder to x[*rd*].

31 25	24 20	19 15	14 12	11 7	6 0
0000001	rs2	rs1	110	rd	0110011

---

remu rd, rs1, rs2 $x[rd] = x[rs1] \%_u x[rs2]$

Remainder, Unsigned. R-type, RV32M and RV64M.
Divides x[*rs1*] by x[*rs2*], rounding towards zero, treating the values as unsigned numbers, and writes the remainder to x[*rd*].

31 25	24 20	19 15	14 12	11 7	6 0
0000001	rs2	rs1	111	rd	0110011

remuw rd, rs1, rs2   x[rd] = sext(x[rs1][31:0] %u x[rs2][31:0])

Remainder Word, Unsigned. R-type, RV64M only.

Divides the lower 32 bits of x[*rs1*] by the lower 32 bits of x[*rs2*], rounding towards zero, treating the values as unsigned numbers, and writes the sign-extended 32-bit remainder to x[*rd*].

31	25 24	20 19	15 14	12 11	7 6	0
0000001	rs2	rs1	111	rd	0111011	

remw rd, rs1, rs2   x[rd] = sext(x[rs1][31:0] %s x[rs2][31:0])

Remainder Word. R-type, RV64M only.

Divides the lower 32 bits of x[*rs1*] by the lower 32 bits of x[*rs2*], rounding towards zero, treating the values as two's complement numbers, and writes the sign-extended 32-bit remainder to x[*rd*].

31	25 24	20 19	15 14	12 11	7 6	0
0000001	rs2	rs1	110	rd	0111011	

ret   pc = x[1]

Return. Pseudoinstruction, RV32I and RV64I.

Returns from a subroutine. Expands to **jalr** x0, 0(x1).

sb rs2, offset(rs1)   M[x[rs1] + sext(offset)] = x[rs2][7:0]

Store Byte. S-type, RV32I and RV64I.

Stores the least-significant byte in register x[*rs2*] to memory at address x[*rs1*] + *sign-extend(offset)*.

31	25 24	20 19	15 14	12 11	7 6	0
offset[11:5]	rs2	rs1	000	offset[4:0]	0100011	

sc.d rd, rs2, (rs1)   x[rd] = StoreConditional64(M[x[rs1]], x[rs2])

Store-Conditional Doubleword. R-type, RV64A only.

Stores the eight bytes in register x[*rs2*] to memory at address x[*rs1*], provided there exists a load reservation on that memory address. Writes 0 to x[*rd*] if the store succeeded, or a nonzero error code otherwise.

31	27 26	25 24	20 19	15 14	12 11	7 6	0
00011	aq	rl	rs2	rs1	011	rd	0101111	

SC.W rd, rs2, (rs1)   x[rd] = StoreConditional32(M[x[rs1]], x[rs2])
Store-Conditional Word. R-type, RV32A and RV64A.
Stores the four bytes in register x[*rs2*] to memory at address x[*rs1*], provided there exists a load reservation on that memory address. Writes 0 to x[*rd*] if the store succeeded, or a nonzero error code otherwise.

31		27 26	25 24		20 19		15 14		12 11		7 6		0
00011		aq	rl	rs2		rs1		010		rd		0101111	

sd rs2, offset(rs1)   M[x[rs1] + sext(offset)] = x[rs2][63:0]
Store Doubleword. S-type, RV64I only.
Stores the eight bytes in register x[*rs2*] to memory at address x[*rs1*] + *sign-extend*(*offset*).
Compressed forms: **c.sdsp** rs2, offset; **c.sd** rs2, offset(rs1)

31		25 24		20 19		15 14		12 11		7 6		0
offset[11:5]		rs2		rs1		011		offset[4:0]		0100011	

seqz rd, rs1   x[rd] = (x[rs1] == 0)
Set if Equal to Zero. Pseudoinstruction, RV32I and RV64I.
Writes 1 to x[*rd*] if x[*rs1*] equals 0, or 0 if not. Expands to **sltiu** rd, rs1, 1.

sext.w rd, rs1   x[rd] = sext(x[rs1][31:0])
Sign-extend Word. Pseudoinstruction, RV64I only.
Reads the lower 32 bits of x[*rs1*], sign-extends them, and writes the result to x[*rd*]. Expands to **addiw** rd, rs1, 0.

sfence.vma rs1, rs2   Fence(Store, AddressTranslation)
Fence Virtual Memory. R-type, RV32I and RV64I privileged architectures.
Orders preceding stores to the page tables with subsequent virtual-address translations. When *rs2*=0, translations for all address spaces are affected; otherwise, only translations for address space identified by x[*rs2*] are ordered. When *rs1*=0, translations for all virtual addresses in the selected address spaces are ordered; otherwise, only translations for the page containing virtual address x[*rs1*] in the selected address spaces are ordered.

31		25 24		20 19		15 14		12 11		7 6		0
0001001		rs2		rs1		000		00000		1110011	

sgtz rd, rs2   x[rd] = (x[rs2] $>_s$ 0)
Set if Greater Than to Zero. Pseudoinstruction, RV32I and RV64I.
Writes 1 to x[*rd*] if x[*rs2*] is greater than 0, or 0 if not. Expands to **slt** rd, x0, rs2.

sh rs2, offset(rs1)   `M[x[rs1] + sext(offset)] = x[rs2][15:0]`

Store Halfword. S-type, RV32I and RV64I.

Stores the two least-significant bytes in register x[*rs2*] to memory at address x[*rs1*] + *sign-extend*(*offset*).

31	25	24	20	19	15	14	12	11	7	6	0
offset[11:5]		rs2		rs1		001		offset[4:0]		0100011	

sw rs2, offset(rs1)   `M[x[rs1] + sext(offset)] = x[rs2][31:0]`

Store Word. S-type, RV32I and RV64I.

Stores the four least-significant bytes in register x[*rs2*] to memory at address x[*rs1*] + *sign-extend*(*offset*).

Compressed forms: **c.swsp** rs2, offset; **c.sw** rs2, offset(rs1)

31	25	24	20	19	15	14	12	11	7	6	0
offset[11:5]		rs2		rs1		010		offset[4:0]		0100011	

sll rd, rs1, rs2   `x[rd] = x[rs1] << x[rs2]`

Shift Left Logical. R-type, RV32I and RV64I.

Shifts register x[*rs1*] left by x[*rs2*] bit positions. The vacated bits are filled with zeros, and the result is written to x[*rd*]. The least-significant five bits of x[*rs2*] (or six bits for RV64I) form the shift amount; the upper bits are ignored.

31	25	24	20	19	15	14	12	11	7	6	0
0000000		rs2		rs1		001		rd		0110011	

slli rd, rs1, shamt   `x[rd] = x[rs1] << shamt`

Shift Left Logical Immediate. I-type, RV32I and RV64I.

Shifts register x[*rs1*] left by *shamt* bit positions. The vacated bits are filled with zeros, and the result is written to x[*rd*]. For RV32I, the instruction is only legal when *shamt*[5]=0.

Compressed form: **c.slli** rd, shamt

31	26	25	20	19	15	14	12	11	7	6	0
000000		shamt		rs1		001		rd		0010011	

slliw rd, rs1, shamt   `x[rd] = sext((x[rs1] << shamt)[31:0])`

Shift Left Logical Word Immediate. I-type, RV64I only.

Shifts x[*rs1*] left by *shamt* bit positions. The vacated bits are filled with zeros, the result is truncated to 32 bits, and the sign-extended 32-bit result is written to x[*rd*]. The instruction is only legal when *shamt*[5]=0.

31	26	25	20	19	15	14	12	11	7	6	0
000000		shamt		rs1		001		rd		0011011	

sllw rd, rs1, rs2

x[rd] = sext((x[rs1] << x[rs2][4:0])[31:0])

Shift Left Logical Word. R-type, RV64I only.

Shifts the lower 32 bits of x[*rs1*] left by x[*rs2*] bit positions. The vacated bits are filled with zeros, and the sign-extended 32-bit result is written to x[*rd*]. The least-significant five bits of x[*rs2*] form the shift amount; the upper bits are ignored.

31	25 24	20 19	15 14	12 11	7 6	0
0000000	rs2	rs1	001	rd	0111011	

slt rd, rs1, rs2

x[rd] = x[rs1] $<_s$ x[rs2]

Set if Less Than. R-type, RV32I and RV64I.

Compares x[*rs1*] and x[*rs2*] as two's complement numbers, and writes 1 to x[*rd*] if x[*rs1*] is smaller, or 0 if not.

31	25 24	20 19	15 14	12 11	7 6	0
0000000	rs2	rs1	010	rd	0110011	

slti rd, rs1, immediate

x[rd] = x[rs1] $<_s$ sext(immediate)

Set if Less Than Immediate. I-type, RV32I and RV64I.

Compares x[*rs1*] and the sign-extended *immediate* as two's complement numbers, and writes 1 to x[*rd*] if x[*rs1*] is smaller, or 0 if not.

31	20 19	15 14	12 11	7 6	0
immediate[11:0]	rs1	010	rd	0010011	

sltiu rd, rs1, immediate

x[rd] = x[rs1] $<_u$ sext(immediate)

Set if Less Than Immediate, Unsigned. I-type, RV32I and RV64I.

Compares x[*rs1*] and the sign-extended *immediate* as unsigned numbers, and writes 1 to x[*rd*] if x[*rs1*] is smaller, or 0 if not.

31	20 19	15 14	12 11	7 6	0
immediate[11:0]	rs1	011	rd	0010011	

sltu rd, rs1, rs2

x[rd] = x[rs1] $<_u$ x[rs2]

Set if Less Than, Unsigned. R-type, RV32I and RV64I.

Compares x[*rs1*] and x[*rs2*] as unsigned numbers, and writes 1 to x[*rd*] if x[*rs1*] is smaller, or 0 if not.

31	25 24	20 19	15 14	12 11	7 6	0
0000000	rs2	rs1	011	rd	0110011	

sltz rd, rs1   x[rd] = (x[rs1] $<_s$ 0)

Set if Less Than to Zero. Pseudoinstruction, RV32I and RV64I.
Writes 1 to x[*rd*] if x[*rs1*] is less than zero, or 0 if not. Expands to **slt** rd, rs1, x0.

snez rd, rs2   x[rd] = (x[rs2] $\neq$ 0)

Set if Not Equal to Zero. Pseudoinstruction, RV32I and RV64I.
Writes 0 to x[*rd*] if x[*rs2*] equals 0, or 1 if not. Expands to **sltu** rd, x0, rs2.

sra rd, rs1, rs2   x[rd] = x[rs1] $>>_s$ x[rs2]

Shift Right Arithmetic. R-type, RV32I and RV64I.
Shifts register x[*rs1*] right by x[*rs2*] bit positions. The vacated bits are filled with copies of
x[*rs1*]'s most-significant bit, and the result is written to x[*rd*]. The least-significant five bits
of x[*rs2*] (or six bits for RV64I) form the shift amount; the upper bits are ignored.

31	25 24	20 19	15 14	12 11	7 6	0
0100000	rs2	rs1	101	rd	0110011	

srai rd, rs1, shamt   x[rd] = x[rs1] $>>_s$ shamt

Shift Right Arithmetic Immediate. I-type, RV32I and RV64I.
Shifts register x[*rs1*] right by *shamt* bit positions. The vacated bits are filled with copies of
x[*rs1*]'s most-significant bit, and the result is written to x[*rd*]. For RV32I, the instruction is
only legal when *shamt*[5]=0.
Compressed form: **c.srai** rd, shamt

31	26 25	20 19	15 14	12 11	7 6	0
010000	shamt	rs1	101	rd	0010011	

sraiw rd, rs1, shamt   x[rd] = sext(x[rs1][31:0] $>>_s$ shamt)

Shift Right Arithmetic Word Immediate. I-type, RV64I only.
Shifts the lower 32 bits of x[*rs1*] right by *shamt* bit positions. The vacated bits are filled with
copies of x[*rs1*][31], and the sign-extended 32-bit result is written to x[*rd*]. The instruction
is only legal when *shamt*[5]=0.

31	26 25	20 19	15 14	12 11	7 6	0
010000	shamt	rs1	101	rd	0011011	

sraw rd, rs1, rs2 $x[rd] = sext(x[rs1][31:0] >>_s x[rs2][4:0])$

Shift Right Arithmetic Word. R-type, RV64I only.

Shifts the lower 32 bits of x[*rs1*] right by x[*rs2*] bit positions. The vacated bits are filled with x[*rs1*][31], and the sign-extended 32-bit result is written to x[*rd*]. The least-significant five bits of x[*rs2*] form the shift amount; the upper bits are ignored.

31	25 24	20 19	15 14	12 11	7 6	0
0100000	rs2	rs1	101	rd	0111011	

sret   ExceptionReturn(Supervisor)

Supervisor-mode Exception Return. R-type, RV32I and RV64I privileged architectures.

Returns from a supervisor-mode exception handler. Sets the *pc* to CSRs[sepc], the privilege mode to CSRs[sstatus].SPP, CSRs[sstatus].SIE to CSRs[sstatus].SPIE, CSRs[sstatus].SPIE to 1, and CSRs[sstatus].SPP to 0.

31	25 24	20 19	15 14	12 11	7 6	0
0001000	00010	00000	000	00000	1110011	

srl rd, rs1, rs2 $x[rd] = x[rs1] >>_u x[rs2]$

Shift Right Logical. R-type, RV32I and RV64I.

Shifts register x[*rs1*] right by x[*rs2*] bit positions. The vacated bits are filled with zeros, and the result is written to x[*rd*]. The least-significant five bits of x[*rs2*] (or six bits for RV64I) form the shift amount; the upper bits are ignored.

31	25 24	20 19	15 14	12 11	7 6	0
0000000	rs2	rs1	101	rd	0110011	

srli rd, rs1, shamt $x[rd] = x[rs1] >>_u shamt$

Shift Right Logical Immediate. I-type, RV32I and RV64I.

Shifts register x[*rs1*] right by *shamt* bit positions. The vacated bits are filled with zeros, and the result is written to x[*rd*]. For RV32I, the instruction is only legal when *shamt*[5]=0.

Compressed form: **c.srli** rd, shamt

31	26 25	20 19	15 14	12 11	7 6	0
000000	shamt	rs1	101	rd	0010011	

srliw rd, rs1, shamt $x[rd] = sext(x[rs1][31:0] >>_u shamt)$

Shift Right Logical Word Immediate. I-type, RV64I only.

Shifts the lower 32 bits of x[*rs1*] right by *shamt* bit positions. The vacated bits are filled with zeros, and the sign-extended 32-bit result is written to x[*rd*]. The instruction is only legal when *shamt*[5]=0.

31	26 25	20 19	15 14	12 11	7 6	0
000000	shamt	rs1	101	rd	0011011	

srlw rd, rs1, rs2 $\qquad$ x[rd] = sext(x[rs1][31:0] >>$_u$ x[rs2][4:0])

Shift Right Logical Word. R-type, RV64I only.

Shifts the lower 32 bits of x[*rs1*] right by x[*rs2*] bit positions. The vacated bits are filled with zeros, and the sign-extended 32-bit result is written to x[*rd*]. The least-significant five bits of x[*rs2*] form the shift amount; the upper bits are ignored.

31	25 24	20 19	15 14	12 11	7 6	0
0000000	rs2	rs1	101	rd	0111011	

sub rd, rs1, rs2 $\qquad$ x[rd] = x[rs1] - x[rs2]

Subtract. R-type, RV32I and RV64I.

Subtracts register x[*rs2*] from register x[*rs1*] and writes the result to x[*rd*]. Arithmetic overflow is ignored.

Compressed form: **c.sub** rd, rs2

31	25 24	20 19	15 14	12 11	7 6	0
0100000	rs2	rs1	000	rd	0110011	

subw rd, rs1, rs2 $\qquad$ x[rd] = sext((x[rs1] - x[rs2])[31:0])

Subtract Word. R-type, RV64I only.

Subtracts register x[*rs2*] from register x[*rs1*], truncates the result to 32 bits, and writes the sign-extended result to x[*rd*]. Arithmetic overflow is ignored.

Compressed form: **c.subw** rd, rs2

31	25 24	20 19	15 14	12 11	7 6	0
0100000	rs2	rs1	000	rd	0111011	

tail symbol $\qquad$ pc = &symbol; clobber x[6]

Tail call. Pseudoinstruction, RV32I and RV64I.

Sets the *pc* to *symbol*, overwriting x[6] in the process. Expands to **auipc** x6, offsetHi then **jalr** x0, offsetLo(x6).

wfi $\qquad$ while (noInterruptsPending) idle

Wait for Interrupt. R-type, RV32I and RV64I privileged architectures.

Idles the processor to save energy if no enabled interrupts are currently pending.

31	25 24	20 19	15 14	12 11	7 6	0
0001000	00101	00000	000	00000	1110011	

xor rd, rs1, rs2   x[rd] = x[rs1] ˆ x[rs2]

Exclusive-OR. R-type, RV32I and RV64I.

Computes the bitwise exclusive-OR of registers x[*rs1*] and x[*rs2*] and writes the result to x[*rd*].

Compressed form: **c.xor** rd, rs2

31	25 24	20 19	15 14	12 11	7 6	0
0000000	rs2	rs1	100	rd	0110011	

xori rd, rs1, immediate   x[rd] = x[rs1] ˆ sext(immediate)

Exclusive-OR Immediate. I-type, RV32I and RV64I.

Computes the bitwise exclusive-OR of the sign-extended *immediate* and register x[*rs1*] and writes the result to x[*rd*].

31	20 19	15 14	12 11	7 6	0
immediate[11:0]	rs1	100	rd	0010011	

B | Transliteration from RISC-V

Plato (428–348 BCE) was a classical Greek philosopher who laid the foundations for Western mathematics, philosophy, and science.

Simplicity

Beauty of style and harmony and grace and good rhythm depend on simplicity.

— Plato, *The Republic.*

B.1 Introduction

This appendix includes tables that transliterate common instructions and idioms in RV32I to equivalent ARM-32 and x86-32 code. Our goal in writing this appendix is to assist programmers who are unfamiliar with RISC-V but are comfortable with ARM-32 or x86-32 to help them learn RISC-V and to help them translate the older ISAs into basic RISC-V code. The appendix concludes with a C routine that traverses a binary tree and with annotated assembly code for all three ISAs. We scheduled the three implementations' instructions as similarly as possible to clarify their correspondence.

The data transfer instructions in Figure B.1 show the similarity between the loads and stores of RV32I and ARM-32 for the most popular addressing mode. Given the memory-register orientation of the x86 ISA instead of the load-store orientation of the RV32I and ARM-32 ISAs, x86 transfers data instead using move instructions.

In addition to the standard integer arithmetic, logical, and shift instructions, Figure B.2 shows how some common operations are done in each ISA. For example, zeroing a register uses the pseudoinstruction li in RV32I, a move immediate instruction in ARM-32, and by exclusive-ORing a register with itself in x86-32. The two operand limit of x86-32 instructions means more instructions in a few cases, although the variable length instruction format lets

Description	RV32I	ARM-32	x86-32
Load word	lw t0, 4(t1)	ldr r0, [r1, #4]	mov eax, [edi+4]
Load halfword unsigned	lh t0, 4(t1)	ldrsh r0, [r1, #4]	movsx eax,WORD PTR[edi+4]
Load halfword	lhu t0, 4(t1)	ldrh r0, [r1, #4]	movzx eax,WORD PTR[edi+4]
Load byte	lb t0, 4(t1)	ldrsb r0, [r1, #4]	movsx eax,BYTE PTR[edi+4]
Load byte unsigned	lbu t0, 4(t1)	ldrb r0, [r1, #4]	movzx eax,BYTE PTR[edi+4]
Store byte	sb t0, 4(t1)	strb r0, [r1, #4]	mov [edi+4], al
Store halfword	sh t0, 4(t1)	strh r0, [r1, #4]	mov [edi+4], ax
Store word	sw t0, 4(t1)	str r0, [r1, #4]	mov [edi+4], eax

Figure B.1: RV32I memory-access instructions transliterated into ARM-32 and x86-32.

Description	RV32I	ARM-32	x86-32
Zero register	li t0, 0	mov r0, #0	xor eax, eax
Move register	mv t0, t1	mov r0, r1	mov eax, edi
Complement register	not t0, t1	mvn r0, r1	not eax, edi
Negate register	neg t0, t1	rsb r0, r1, #0	mov eax, edi neg eax
Load large constant	lui t0, 0xABCDE addi t0, t0, 0x123	movw r0, #0xE123 movt r0, #0xABCD	mov eax, 0xABCDE123
Move PC to register	auipc t0, 0	ldr r0, [pc, #-8]	call 1f 1: pop eax
Add	add t0, t1, t2	add r0, r1, r2	lea eax, [edi+esi]
Add (imm.)	addi t0, t0, 1	add r0, r0, #1	add eax, 1
Subtract	sub t0, t0, t1	sub r0, r0, r1	sub eax, edi
Set reg. to (reg=0)	sltiu t0, t1, 1	rsbs r0, r1, #1 movcc r0, #0	xor eax, eax test edx, edx sete al
Set reg. to (reg≠0)	sltu t0, x0, t1	adds r0, r1, #0 movne r0, #1	xor eax, eax test edx, edx setne al
Bitwise-OR	or t0, t0, t1	orr r0, r0, r1	or eax, edi
Bitwise-AND	and t0, t0, t1	and r0, r0, r1	and eax, edi
Bitwise-XOR	xor t0, t0, t1	eor r0, r0, r1	xor eax, edi
Bitwise-OR (imm.)	ori t0, t0, 1	orr r0, r0, #1	or eax, 1
Bitwise-AND (imm.)	andi t0, t0, 1	and r0, r0, #1	and eax, 1
Bitwise-XOR (imm.)	xori t0, t0, 1	eor r0, r0, #1	xor eax, 1
Shift left	sll t0, t0, t1	lsl r0, r0, r1	sal eax, cl
Shift right logical	srl t0, t0, t1	lsr r0, r0, r1	shr eax, cl
Shift right arith.	sra t0, t0, t1	asr r0, r0, r1	sar eax, cl
Shift left (imm.)	slli t0, t0, 1	lsl r0, r0, #1	sal eax, 1
Shift right logical (imm.)	srli t0, t0, 1	lsr r0, r0, #1	shr eax, 1
Shift right arith. (imm.)	srai t0, t0, 1	asr r0, r0, #1	sar eax, 1

Figure B.2: RV32I arithmetic instructions transliterated into ARM-32 and x86-32. The two-operand x86-32 instruction format often needs more instructions than the three-operand instruction format of ARM-32 and RV32I.

Description	RV32I	ARM-32	x86-32
Branch if $=$	beq t0, t1, foo	cmp r0, r1 beq foo	cmp eax, esi je foo
Branch if $\neq$	bne t0, t1, foo	cmp r0, r1 bne foo	cmp eax, esi jne foo
Branch if $<$	blt t0, t1, foo	cmp r0, r1 blt foo	cmp eax, esi jl foo
Branch if $\geq_s$	bge t0, t1, foo	cmp r0, r1 bge foo	cmp eax, esi jge foo
Branch if $<_u$	bltu t0, t1, foo	cmp r0, r1 bcc foo	cmp eax, esi jb foo
Branch if $\geq_u$	bgeu t0, t1, foo	cmp r0, r1 bcs foo	cmp eax, esi jnb foo
Branch if $=0$	beqz t0, foo	cmp r0, #0 beq foo	test eax, eax je foo
Branch if $\neq0$	bnez t0, foo	cmp r0, #0 bne foo	test eax, eax jne foo
Direct jump or tail call	jal x0, foo	b foo	jmp foo
Subroutine call	jal ra, foo	bl foo	call foo
Subroutine return	jalr x0, 0(ra)	bx lr	ret
Indirect call	jalr ra, 0(t0)	blx r0	call eax
Indirect jump or tail call	jalr x0, 0(t0)	bx r0	jmp eax

Figure B.3: RV32I control-flow instructions transliterated into ARM-32 and x86-32. The compare-and-branch instruction of RV32I takes half the number of instructions of the condition code based branches of ARM-32 and x86-32.

it load a large constant in a single instruction. The conventional add, subtract, logical, and shift instructions—which are responsible for most instructions executed—map one-to-one between ISAs.

Figure B.3 lists the conditional and unconditional branch and call instructions. The condition code approach to conditional branches requires two instructions for ARM-32 and x86-32 while RV32I needs just one. As Chapter 2 illustrates in Figures 2.5 to 2.11, despite its minimalist approach to instruction set design, the compare-and-execute branches of RISC-V shrink the number of instructions in Insertion Sort by as much as the fancier address modes and the push and pop instructions of ARM-32 and x86-32.

Performance

B.2 Comparing RV32I, ARM-32, and x86-32 using Tree Sum

Figure B.4 is our example C program that we use to compare the three ISAs side-by-side in Figures B.5 to B.7. It sums the values in a binary tree, using an in-order tree traversal. Trees are a fundamental data structure, and while this tree operation might seem overly simplistic, we picked it because it demonstrates both recursion and iteration in only a few assembly instructions. The routine recurses to compute the sum of the left subtree, but uses iteration to compute the sum of the right subtree, which reduces the memory footprint and instruction count. Optimizing compilers can transform the fully recursive code into this version; we show the iteration explicitly for clarity.

The biggest differences between the size of the three assembly language programs is in the function entry and exit. RISC-V uses four instructions to save and restore three registers on the stack and to adjust the stack pointer. x86-32 saves and restores only two registers on the stack because it can perform arithmetic operations on memory operands instead of loading them all into registers. It also saves and restores them using push and pop instructions, which implicitly adjusts the stack pointer rather than doing it explicitly as in RISC-V. ARM-32 can save three registers plus the link register with the return address on the stack in a single push instruction, and restore them with a single pop instruction.

RISC-V executes the main loop itself in seven instructions instead of eight for the other ISAs because, as Figure B.3 shows, it can compare-and-branch in a single instruction while that operation takes two instructions for ARM-32 and x86-32. The rest of the instructions in the loop map one-to-one between RV32I and ARM-32, as Figures B.1 and B.2 illustrate. One difference is that x86's `call` and `ret` instructions implicitly push and pop the return address to and from the stack, whereas the other ISAs instead do so explicitly in their prologues and epilogues (by saving and restoring `ra` for RV32I, or by pushing `lr` and popping to `pc` for ARM-32). Also, because the x86-32 calling convention passes arguments on the stack, the x86-32 code has a `push` and a `pop` instruction in the loop the other ISAs can avoid. The extra data transfer reduces performance.

Performance

B.3   Conclusion

Despite widely different ISA philosophies, the resulting programs are quite similar, making it straightforward to translate from the versions of the program of the older architectures to RISC-V. Having 32 registers for RISC-V versus 16 for ARM-32 and 8 for x86-32 simplifies translating to RISC-V, which would be much harder in the other direction. First adjust function prologues and epilogues, then change the conditional branches from condition code oriented to compare-and-branch instructions, and finally replace all the register and instruction names with the RISC-V equivalents. There might be a few more adjustments remaining, such as handling long constants and addresses in the variable-length x86-32 ISA or adding RISC-V instructions to accomplish the fancy addressing modes if used in data transfers, but you would be close after following only these three steps.

Programmability

```c
struct tree_node {
  struct tree_node *left;
  struct tree_node *right;
  long value;
};

long tree_sum(const struct tree_node *node)
{
  long result = 0;
  while (node) {
    result += tree_sum(node->left);
    result += node->value;
    node = node->right;
  }
  return result;
}
```

Figure B.4: A C routine that sums the values in a binary tree, using an in-order traversal.

```
        addi sp,sp,-16   # Allocate stack frame
        sw   s1,4(sp)    # Preserve s1
        sw   s0,8(sp)    # Preserve s0
        sw   ra,12(sp)   # Preserve ra
        li   s1,0        # sum = 0
        beqz a0,.L1      # Skip loop if node == 0
        mv   s0,a0       # s0 = node
.L3:
        lw   a0,0(s0)    # a0 = node->left
        jal  tree_sum    # Recurse; result in a0
        lw   a5,8(s0)    # a5 = node->value
        lw   s0,4(s0)    # node = node->right
        add  s1,a0,s1    # sum += a0
        add  s1,s1,a5    # sum += a5
        bnez s0,.L3      # Loop if node != 0
.L1:
        mv   a0,s1       # Return sum in a0
        lw   s1,4(sp)    # Restore s1
        lw   s0,8(sp)    # Restore s0
        lw   ra,12(sp)   # Restore ra
        addi sp,sp,16    # Deallocate stack frame
        ret              # Return
```

Figure B.5: RV32I code for in-order tree traversal. The main loop is shorter than the versions for the other two ISAs due to the compare-and-branch instruction bnez.

```
      push {r4, r5, r6, lr}  # Preserve regs
      mov  r5, #0            # sum = 0
      subs r4, r0, #0        # r4 = node; node == 0?
      beq  .L1               # Skip loop if so
  .L3:
      ldr  r0, [r4]          # r0 = node->left
      bl   tree_sum          # Recurse; result in r0
      ldr  r3, [r4, #8]      # r3 = node->value
      ldr  r4, [r4, #4]      # r4 = node->right
      add  r5, r0, r5        # sum += r0
      add  r5, r5, r3        # sum += r3
      cmp  r4, #0            # node == 0?
      bne  .L3               # Loop if not
  .L1:
      mov  r0, r5            # Return sum in r0
      pop  {r4, r5, r6, pc}  # Restore regs and return
```

Figure B.6: ARM-32 code for in-order tree traversal. The multiword push and pop instructions reduce the code size for ARM-32 versus the other ISAs.

```
    push esi            # Preserve esi
    push ebx            # Preserve ebx
    xor  esi, esi       # sum = 0
    mov  ebx, [esp+12]  # ebx = node
    test ebx, ebx       # node == 0?
    je   .L1            # Skip if so
  .L3:
    push [ebx]          # Load node->left; push to stack
    call tree_sum       # Recurse; result in eax
    pop  edx            # Pop old arg and discard
    add  esi, [ebx+8]   # sum += node->value
    mov  ebx, [ebx+4]   # node = node->right
    add  esi, eax       # sum += eax
    test ebx, ebx       # node == 0?
    jne  .L3            # Loop if not
  .L1:
    mov  eax, esi       # Return sum in eax
    pop  ebx            # Restore ebx
    pop  esi            # Restore esi
    ret                 # Return
```

Figure B.7: x86-32 code for in-order tree traversal. The main loop has push and pop instructions not found in the versions of the program for the other ISAs, which create extra data traffic.

Index